THE UFO, E.T, ALIEN TRILOGY

Martin Thomas

THE UFO, E.T, ALIEN TRILOGY

The moral rights of the author have been asserted by them.

© Martin Thomas 2025

10 9 8 7 6 5 4 3 2 1

This version has an index and a list of references, as well as some amendments.

All rights reserved

No part of this publication may be reproduced, stored in a retrieval system, or transmitted in any form or by any other means (electronic, mechanical, photocopying, recording or otherwise) or used to train any artificial intelligence technologies without the prior written permission of the author. Subject to EU law, the author expressly reserves this work from the text and data mining exemption.

Previous Publications by the Author:

UFO – Friend or Foe?	August 2025	978-1-918045-21-5
We Own 29% - ET Has the Rest	September 2025	978-1-83709-256-7
Earth's Alien Syllabus	October 2025	978-1-83709-274-1

Web Address: MartinThomasAuthor.com

Individual copies of the three publications in this book can still be acquired via my website.

Dedication

This work is dedicated to the BLAZE Television Channel for making me think, to Google for helping me find information and to Wikipedia for denying everything.

The Scientific Method

Develop an hypothesis which can be used to make testable predictions. Test these predictions and if they work, test again for repeatability.

The Engineering Method

If it walks like a duck, and quacks like a duck, it probably is a duck.

THE UFO, E.T, ALIEN TRILOGY

THE UFO, E.T, ALIEN TRILOGY

Table of Contents

    Preface   vii

PART ONE – UFO – Friend or Foe?   1

    Table of Contents for Part One   3

    Preface to Part One   5

1   IF THEY AREN'T "OTHER" IN ORIGIN, WHAT ARE UAPs ?   7

2   WHAT ACTIVITIES ARE ATTRIBUTED TO THE "OTHERS"?   14

3   SO WHAT ARE THESE "OTHERS"?   52

4   WHERE ARE THEY?   63

5   WHY IS THERE NO NATIONAL OR INTERNATIONAL RESPONSE?   71

6   OUR GUARDIANS   79

PART TWO – We Own 29% - ET Has the Rest   81

    Table of Contents for Part Two   83

    Preface to Part Two   85

1   INTRODUCTION   87

2   WHERE DID INTER-TERRESTRIALS COME FROM?   93

3   WHERE ARE THE MOST LIKELY BASES?   98

4   WHAT "OTHERS" DO WE KNOW ABOUT?   140

5   WHERE ARE THEY HIDING?   153

6   CONCLUSIONS   160

    APPENDIX   162

PART 3 – EARTH'S ALIEN SYLLABUS — 203

    Table of Contents For part 3 — 205

    Preface for Part 3 — 207

1. THE 50,000 YEAR PLAN — 209
2. 40,000 YEARS LATER — 212
3. THE LAST THOUSAND YEARS — 221
4. THE LAST CENTURY — 223
5. THE WAY FORWARD — 245

    APPENDICES — 247

PART 4 - INDEX & REFERENCES — 287

    INDEX — 289

    REFERENCES — 298

PREFACE

Millions of people believe that there are mysterious objects flying in the sky. They were originally described as Unidentified Flying Objects – hence UFOs – but have more recently been termed UAPs – Unidentified Anomalous Phenomena. As most humans spend their time on the land, there are many more sightings of UAPs there, than have been associated with maritime UFOs, which have been termed Unidentified Submersible Objects (USOs). The abbreviation UAP is now used to cover both UFOs and USOs.

The term "Alien" has been used to describe any occupants or manufacturers of UAPs. However, this word has many pejorative associations – it means basically "not one of us" and is frequently used to describe immigrants, whose presence may be illegal, or who are disliked because of their race, colour or creed. I intend to use the word "Others", perhaps not an ideal term, but at least it avoids many of the pre-assumptions inherent in traditional terms.

THE UFO, E.T, ALIEN TRILOGY

PART 1

UFO
FRIEND OR FOE?

THE UFO, E.T, ALIEN TRILOGY

THE UFO, E.T, ALIEN TRILOGY

Table of Contents

    Preface   5

1. **IF THEY AREN'T "OTHER" IN ORIGIN, WHAT ARE UAPs ?**   7
   - a) GENERAL   7
   - b) THEY DON'T EXIST   8
   - c) THEY ARE HOAXES   8
   - d) THEY ARE MISS-INTERPRETED NATURAL PHENOMENA   9
   - e) MULTIPLE REPORTS ARE MASS HYSTERIA   10
   - f) THEY ARE PRODUCTS OF ADVANCED EARTHLY TECHNOLOGY   10

2. **WHAT ACTIVITIES ARE ATTRIBUTED TO THE "OTHERS"?**   14
   - a) PRE-HUMANITY   14
   - b) EARLY HUMANITY TO WORLD WAR II   15
   - c) WORLD WAR II AND BEYOND   21
     - i   General   21
     - ii   USA   22
     - iii Canada   28
     - iv USSR/Russia   29
     - v   United Kingdom   31
     - vi Mainland Europe   34
     - vii Central America and the Caribbean   36
     - viii South America   38
     - ix Japan and the Dragon's Triangle   41
     - x   Australia & New Zealand   42
     - xi World-Wide   47
     - xii Summary   48
   - d) WHEN THEY MAKE CONTACT   49
     - i   Contact with the Authorities   49
     - ii Contact with Real People   49
     - iii What to believe?   51

3. **SO WHAT ARE THESE "OTHERS"?**   52
   - a) INTRODUCTION   52
   - b) THEY ARE CAUSED BY GODS OR MAGICIANS   52
   - c) THEY ARE FROM OUR FUTURE   53
   - d) THEY ARE FROM AN EARLIER HUMAN CIVILISATION IN HIDING   53

	e)	APES WEREN'T THE ONLY EARTHLY CREATURES TO EVOLVE	54
	f)	THEY COME FROM PARALLEL UNIVERSES	55
	g)	THEY ARE PRODUCTS OF AN "OTHER" CULTURE	57
		i General	57
		ii Intra-Terrestrials (ITs)	58
		iii Extra-Terrestrial (ETs)	60
	h)	THEY ARE HUMAN PRODUCTS OF REVERSE ENGINEERING OF "OTHER" TECHNOLOGY	61
4	**WHERE ARE THEY?**		63
	a)	WHERE ARE THEIR BASES?	63
		i Elsewhere	63
		ii Underground	63
		iii Under water	66
		iv Possible Bases Elsewhere	68
	b)	THE BALD FACTS	68
	c)	COLONISATION IN SPACE	70
5	**WHY IS THERE NO NATIONAL OR INTERNATIONAL RESPONSE?**		71
	a)	THE USA & ETs	71
	b)	POSSIBLE "OTHER" INTERESTS	72
	c)	MAJESTIC'S ACTIVITIES	73
	d)	RECENT EVENTS IN THE USA SPACE PROGRAMME	75
	e)	THE PRESENT SITUATION	77
6	**OUR GUARDIANS**		79
	a)	GENERAL	79
	b)	AERIAL BATTLES	79
	c)	SUMMARY	80

PREFACE

Physicists, working in the field of String Theory, seem happy to contend with multiple universes, perhaps even an infinite number of them. It may be possible to move via "Portals" from one universe to another or, via "Gateways" into another universe and back into our universe again somewhere else. I am deliberately avoiding the term "Wormhole" which is another term for "Einstein-Rosen Bridge" which postulates links within our universe from one point to another, but which may be too dangerous to use anywhere near a planet.

Native American Shamans suggest that they can open Portals and communicate with the occupants of other universes, making use of vibrations – particularly drums and chanting. Shamans have been filmed inviting small UAPs to appear, and their consequent arrival, on the TV programme "Beyond Skinwalker Ranch". In the Indian sub-continent, the use of chanting appears to serve the same purpose, with mantras claimed to be very powerful.

For Gateways to work, the natures of Portals and Gateways may need to be fundamentally different. To transfer between universes may not require movement on the other side. To come out somewhere else in our universe may.

The obvious questions are: where have these UAPs come from, and what are they doing? It is not my intention to investigate the second of these questions, beyond where it may impact on the first one. An answer to this first question may assist in determining the answer to the second.

It must be remembered that we have no idea of the motivations of these "Others". They could be benign, helping our development as a species from pure altruism. They may want something from us or our planet, but are prepared to make deals with us. They may want our planet. Given the large number of "Other" species in the universe[i] all are possible at the same time. We have to hope some of these species are protecting us.

Chapter 1

IF THEY AREN'T "OTHER" IN ORIGIN, WHAT ARE UAPs ?

a) <u>GENERAL</u>

UAPs seem to fall mainly into 2 distinct categories. They may be objects which are apparently metallic in construction, of many shapes and which can vary in size from larger than a car to being absolutely massive. Alternatively they can be balls, either of some form of plasma, or metallic, giving off or reflecting light, and varying from the size of a tennis ball upwards.

Let's look at the various explanations which have been put forward for where these UAPs come from, but which don't require physics which is currently beyond our comprehension:

- They don't exist
- They are hoaxes
- They are miss-interpreted natural phenomena
- Multiple reports are mass hysteria
- They are products of advanced earthly technology

Some governments make use of such explanations to refute possible UAP sightings, seeming to go out of their way to discredit witnesses, offering patently unscientific explanations, or ignoring inconvenient facts.

So, let's tackle these various proposals one by one.

b) THEY DON'T EXIST

There are always people who chose to deny any evidence, even that of their own eyes. These Doubting Thomases don't feel the need to offer any explanation for what many people describe, and so don't deserve to have the benefit of being taken seriously themselves. Many governments appear to be determined to convince the general population that there is no such thing as a UAP, even to the extent of rubbishing sincere eye-witnesses.

c) THEY ARE HOAXES

Undoubtedly there are many modern day hoaxes which do not survive proper examination. The TV series "The Proof is Out There" does a good job of testing these. Nevertheless, there are sightings which cannot be explained. Generally hoaxers relied on models and strings and, more recently, on doctored photographs and video recordings. However, multiple eye-witnessed events in the sky are sometimes difficult to explain.

There are also images from the past such as mediaeval paintings with apparent UAPs, such as "The Madonna with Saint Giovannino" attributed to Ghirlandaio, and "The Annunciation with Saint Emidius" by Carlo Crivelli". There is also a woodcut dated 1561 of a mass sighting of celestial objects over Nuremberg[2] which is difficult to ignore.

There are also anachronous artefacts found in genuine ancient archaeological strata. Examples might be the Antikythera[3] Mechanism, the Coso Artifact[4] and the Palenque Astronaut[5].

Certainly, not all UAPs can be hoaxes.

d) <u>THEY ARE MISS-INTERPRETED NATURAL PHENOMENA</u>

Many UAPs sightings are mistakes. Recording Venus as a UAP is reputedly a common error, and sightings of high-level American spy planes were apparently frequent (even if not a natural phenomenon). Lenticular[6] clouds, in particular, can be mistaken for massive "Mother Ships". It is theorised that balls of plasma can be generated by some earthquakes and ball lightening is an accepted short term phenomenon. These have all been government options for debunking reported sightings, with little respect for scientific credibility. The UK proposed the exact coincidence of a meteorite and an earthquake in a non-earthquake zone, to explain one sighting[7]. The USA once chose to explain a high-level ball of light as marsh gas[8].

These possible miss-interpretations do not mean that every UAP report is false. It has been estimated that, even with an honest debunking of many UAP reports, about 5% remain unexplained. By excluding sightings of solitary lights in the sky, I hope that this percentage is increased.

e) MULTIPLE REPORTS ARE MASS HYSTERIA

This has been claimed for at least four multiple sightings by schoolchildren in Broadhaven[9] in Wales, Melbourne[10] in Australia, Miami[11] in the USA and Ruwa[12] in Zimbabwe. In all four cases, the witnesses, who are now adult, still stand by their descriptions of what they saw. It is also convenient for the nay-sayers to forget that their teachers were witnesses too. It has been reported that these teachers have been threatened with destruction of their careers if they don't keep silent.

The residents of Phoenix, Arizona would have been greatly offended if the sightings over their city in 1997 were described as mass hysteria, particularly as they were seen by an estimated 10,000 citizens. The event is described as the Phoenix Lights[13], and it was quickly explained away by the military as a flight of military aircraft and other aircraft dropping flares. The Arizona state governor, Fife Symington, responded with humour, introducing a "captured alien" who was actually his chief of staff, in an alien costume. However, once he had retired in 2007, he stated that he had acted in that way to prevent panic, and that he had actually seen the lights and, being an ex-pilot, knew they were neither conventional military aircraft nor flares.

f) THEY ARE THE PRODUCTS OF ADVANCED EARTHLY TECHNOLOGY

Many of the witnesses of small aerial UAPs[14] have been highly qualified pilots, both commercial and military. Their testimonies have been uniformly the same – "We can't manoeuvre like that". These UAPs are reported as showing instantaneous acceleration and deceleration to a far higher degree than human bodies could withstand, and sometimes jumping from one place to another. Also, if any nation had developed this technology, and it would have to be a major state to be able to afford it, its progressive development would have led to earlier applications which would probably have been used in a military context by now.

The larger UAPs seem to have been around for millennia: flying saucers of different sizes, big cylindrical vessels and massive "mother ships". They exceed the speed of sound without a sonic boom, and some have been seen to transition from air to water without apparently displacing water.

Circular UAPs have also been observed for millennia, ranging from 20 feet (7m) across, upwards to what have been called mother-ships. These could not be the product of current technology, as they easily pre-date it. However, in 1967, Jack Picket[15], a publisher of USA Air Force Publications, states that he was shown a range of circular aircraft at MacDill Air Force Base in Tampa which were dumped in their salvage yard. He was shown photographs of them in flight. These were up to 100 feet (30m) in diameter and were jet-powered. He was told that they had been very fast, and could fly very high, but had manoeuvrability problems.

There have been multiple reports of large triangular UAPs the size of football fields, in the Hudson Valley[16] in the USA in the 1980s, in the so-called Belgian Wave[17] in the early 1990s, in Bakewell[18] in the UK in 1993, in the Phoenix Lights incident in 1997, the Southern Illinois Incident[19] in 2000, the Tinley Park Lights[20] in Illinois in 2004-6, and the mass sighting at Stephensville Texas[21] in 2008.

It is not clear whether these are truly triangular, or boomerang shaped, or perhaps both types exist, but they all exhibit light sources on the underside which appear to be the method of levitation and drive. Reports of these sightings suggest that they can hover and fly very rapidly, possibly turning by rotating on the spot, are effectively silent apart from a gentle hum, leave no discernible exhaust, and don't appear to work on current aerodynamic principles.

The location of these sightings points the finger of suspicion at the USA, but this doesn't explain how the technology was developed without some hint of what was going on. There certainly has been a determined attempt in the USA to discredit sightings and witnesses, with scary tales of thugs, sometimes called "Men in Black[22]", turning up to threaten some witnesses and even of sudden fatalities.

Many nations are developing drones technology, with perhaps the USA, Russia, Ukraine and China at the forefront. There are many reports of small bright lights and small solid objects. Many sightings appear to have been in East Asia and South America, with military videos of sightings on test ranges in the USA. There have been videos of small illuminated UAPs appearing and disappearing. It is not clear why Human military drones should even have the capability to be brightly illuminated, when invisibility would appear to the preferable.

Whilst modern drones can do amazing manoeuvres because they don't have pilots to limit the stresses they can take, it is doubtful whether even they can do everything reported, and there have been reports of highly manoeuvrable spherical UAPs from long before the electronic age when human drones first appeared.

In early South American cultures, there are detailed carvings of space beings[23] and spaceships. There are even early miniatures of aircraft[24] which have been shown experimentally to have credible aerodynamic capabilities. It is interesting that, in many places in South America, where ancient structures still stand, the local inhabitants claim that these buildings were already there when their ancestors first came to the region. Skulls[25] have been found in caves there, which are elongated at the back, similar to representations of Egyptian pharaohs such as Akhenaten. Mainstream archaeologists claim that these result from head-binding babies, but it has been shown that this does not enlarge the brain, whereas some of these ancient skulls have significantly larger brain volumes[26].

Chapter 2

WHAT ACTIVITIES ARE ATTRIBUTED TO THE "OTHERS"?

a)   PRE-HUMANITY

No-one seems interested in going back earlier than the age of the dinosaurs but, in some quarters, the "Others" get the blame for launching the asteroid which wiped them out. Unless they tell us that they did, we will never know.

Some geneticists believe that there is insufficient time between the extinction of the dinosaurs 66 million years ago, and the rise of humanity 20 thousand years ago, for natural selection to achieve what it did to humans. Other mammals in the fossil record appear to advance at a predictable rate, but humanity shows remarkable changes in the brain, which may have happened far too quickly. It is suggested there is a genetic discontinuity[27], which was engineered by the "Others".

b)   EARLY HUMANITY TO WORLD WAR II

Of all the achievements credited to the "Others", the most famous is the construction of the huge number of stone buildings from the

Mesolithic period onwards, ranging from the stone circles in Orkney and Stonehenge in the UK, via Nan Madol[28] in the Pacific and the Floating Rock[29] in Japan, to Puma Puncu[30] in Bolivia. In these places, and many more, there are carved stones so massive that it seems impossible that they were the handiwork of humans to carve them or move them. Many of these stones are granite, suggesting that the quartz crystals within could be used for some electrical or magnetic effect.

Early humans were hunter-gatherers until "Others" landed and became their leaders, teachers or kings. Egypt's pharaoh, Akhenaten[31], father of Tutankhamun, appears to have a misshapen head as did his wife and son, with it enlarged to the back and with other non-human differences. It is widely suggested that they were "Other" in origin. Similar misshapen skulls have been found in Malta and South America. Human parents, presumably wishing to make their children appear of royal background, have bound their babies' heads to achieve the same shape. However, it has been shown that they retain the same cranial volume as other children, whilst genuine misshaped skulls are much larger in volume.

There are records suggesting that the Sumarian civilisation was visited by winged "Others" which they called Anunnaki[32]. These beings were supposed to be humanoids, at least 4m tall, perhaps winged, and whose planet Nibiru is on a 3600 year orbit around our sun. They gave the early Sumerians agriculture, reading and writing, the wheel, and a code of law. Similar beings are represented on the walls of Egyptian buildings. The Anunnaki are sometimes shown accompanied by 2.5m tall humanoid beings with bird-like heads called Anzu in Sumerian legends[33], and Horus in Egypt[34]. These sound similar to bird-like "Others" claimed to live in the northern mountains of Japan today.

In the Book of Enoch, which is included in some versions of the Bible, he describes his being taken up into a spaceship where he meets many "Others". He is reputed to be the great grandfather of Noah.

China's first Emperor Qui Shi Huang[35] is thought to have either been "Other" or closely assisted by "Others". In Mexico, their Gods Quezalcoatl[36], Viracotcha[37] and many others are frequently described as the beings who came from the sky and taught the natives the basics of agriculture, medicine, mathematics and building.

In India, the "Others" appeared to have landed mob-handed, with spaceships called Vimanas[38], roughly conical UAPs, stylised representations of which are today a common feature in South East Asian architecture. In some early Indian drawings[39] they are shown as blue-skinned and humanoid in stature so they could be from Vega. They helped local humans but also squabbled amongst themselves. This squabbling culminated in the Kurushekta War[40] in about 1000 BCE, where Indian lore describes the gods flying round, and the use of devices to throw fire like lasers. Archaeologists have found the city Mohenjo Daro[41] to have been completely flattened, and there are claims that it was covered in a layer of radioactive[42] ash, suggesting the use of a nuclear weapon.

Indian history suggests that they were in fact visited by two species of god – the Devas and the Asuras. These latter can be translated as Demons or Titans. Greek folklore describes a war between the Olympians and the Titans, won by the Olympians. The Olympians were human in appearance, whilst the Titans were giants. In India, the Titans are pictured as giants in their carvings, so they could be the same "Other" species. The winners write the history, so perhaps the Titans weren't quite as bad as they are pictured. Certainly, their presence would have assisted in many of the grand construction projects of those times.

Located in the in the fertile basin of the Upper Viliuy River, in Siberia, there is an area known to the nomad Siberian people as the Valley of Death. Here everything is so poisoned that nothing can live there. Across Siberia, giant metal structures may be found, hollow and half-buried in the soil of the forest. These have been nicknamed "Cauldrons[43]". Thanks to the remoteness of the region, their existence was thought to be a myth for many years. The entire area has always been shrouded with mystery, but a few facts emerge from history to tantalize modern scientists. The people of the area remember a legend that tells of a pillar of light. That light shone for many nights, and finally dimmed, only to flare again. It sent a fireball straight up into the air, which then sped across the land to strike down and obliterate the village's enemies. This occurred many times, sometimes in quick succession, and sometimes only after a wait of more than a hundred years.

Was this some form of world defence system? Are there similar "Cauldrons" elsewhere on earth, hidden in deserted areas such as the high Andes, Alaska and central Australia?

It is claimed that the "Others" either built or caused the building of pyramids, not just in Egypt, but everywhere in South and Central America, and in China and South East Asia. These are often associated with obelisks. It is suggested that these together once contributed to a world-wide power grid, with the Great Pyramid of

Khufu being a prime source[44]. It is interesting to note that the obelisks and many menhirs are carved from granite with piezo-electric capabilities from the quartz in it.

The three pyramids at Giza are each aligned to due north, and it is claimed that they are aligned the same way as Orion's Belt[45]. The pyramids at Teotihuacan[46] in Mexico, specifically the Pyramid of the Moon, the Pyramid of the Sun, and the Pyramid of Quetzalcoatl[47], are often claimed to align with Orion's Belt.

The site of Göbekli Tepe[48] in Turkey has been proved to be over 11,000 years old, confounding archaeology's accepted time-line. The statues there show figures with six fingers.

There are now submerged coastal cities which show how the sea-level has risen since the end of the ice-age. As all the water held in the ice melted, it could have caused massive floods, and raised the sea level. Although the great flood, which is recorded in the Bible and other east Mediterranean texts, is generally assumed to be the Mediterranean Sea breaking into the Black Sea, there are records of a Great Flood in many cultures around the world. Some of these cultures are based in relatively high ground, and not all the melt-water in the world could produce floods at 3800m in the Andes or on a 1000m high plateau in Cappadocia[49]. It has been suggested that all the water came from a bombardment of ice-based planetessimals but, if there was sufficient water to cover all or most of earth, where did it all go?

It is suggested that there is evidence of flooding at Puma Puncu[50] in Bolivia in the Andes at a height of 12,000 feet, but this doesn't have to have come from a massive rise in sea level. It is very close to

Lake Halnaymarca, and there may have been something to shake that lake at some time.

In 312, the Emperor Constantine saw a cross in the sky before the battle of Milvian Bridge, and this convinced him to convert to Christianity[51]. It has been suggested that as this occurred in the vicinity of Monte Musine, it might have been an UAP.

The UK had its first recorded UAP sighting in 1113[52], when pilgrims in the SW of England reported seeing a fire-belching dragon emerge from the sea, fly into the air and disappear into the sky.

In 1317, a green circular UAP[53] hung over the Russian city of Tver, emitting a red glow and three rays. It was there for over a week before moving off.

At around dawn on 14 April 1561, according to a broadsheet published that same year, "many men and women" of Nuremberg[54], Germany, saw what the broadsheet describes as an aerial battle with UAPs coming "out of the sun", followed by the appearance of a large black triangular object. Exhausted combatant spheres fell to earth in clouds of smoke. The broadsheet claims that witnesses observed hundreds of spheres, cylinders, and other odd-shaped objects that moved erratically overhead. The woodcut illustration depicts objects of various shapes, including crosses (with or without spheres on the arms), small spheres, two large crescents, a black spear, and cylindrical objects from which several small spheres emerged and darted around the sky.

In April 1665, six fishermen claimed to witness an unexplained celestial phenomenon – an aerial battle in the skies above the Baltic Sea near Stralsund[55]. As evening broke, a dark-gray disk appeared high above the city centre. Great flocks of birds in the sky morphed into warships and engaged in a thunderous air battle. The decks teemed with ghostly figures. When, at dusk, "a flat, round shape like a plate" appeared above the local St. Nicholas Church, they fled. The following day, the fishermen found that they were trembling all over and complained of pains.

On June 30th 1908, in Tunguska[56], Siberia, an oval-shaped mass swept across the sky and then there was a huge explosion. Locals reported that many trees were blown flat and more continued blazing for weeks. It was thought to have been a meteorite 200 feet in diameter, weighing some 100,000 tons. There were several expeditions to find it, between 1927 and 1939, without finding any crash crater. Various theories are current: it was the crash, take-off or landing of a spaceship, it was a test of Tesla's secret energy weapon, or it was a signal from another species. No-one knows, but there are reports that some locals showed signs of genetic mutations similar to those caused by radiation. A further expedition in 1960 found quantities of nickel and iridium, suggesting there had been a mid-air explosion, and globules of melted dust suggested a nuclear event. An expedition in 2004 claimed to have found an "extraterrestrial device", but its nature has not been disclosed.

An UAP is reputed to have crash-landed in Italy in Magenta[57] in 1933. It was supposedly examined by Marconi at the time, and taken away by the USA after the war.

In 1936, an UAP allegedly crashed in the Black Forest near Frieberg[58], Germany and Hitler's SS quickly recovered it for research. It is suggested it was used as the basis of a Nazi flying saucer or Time Machine. However, there appears to be no evidence to support this until almost 30 years later.

c) WORLD WAR II AND BEYOND

i. General

During the Second World War, both Allied and Axis pilots reported what became known as Foo Fighters[59]. These were balls of light which were capable of out-manoeuvring their planes, and could interfere with their electronics if they got too close. There are even reports of their interfering with their engine electrical systems which were generally more low-tech than say radar systems. Allied pilots thought these were Axis secret weapons, and vice-versa.

There were several reports of larger flying saucers during the war, both in the USA and the UK. It is reported[60] that Winston Churchill and General Eisenhower discussed how to deal with the encounter, agreeing on a cover-up.

There are many reports of "Other" sightings during the large-scale NATO exercise Operation Mainbrace[61] in 1953. One of the participating warships was USS Eisenhower, which was the first USA ship to carry nuclear weapons, and was already no stranger to UAPs

Since World War II, the world has changed beyond measure, with human-produced rockets and the arrival of the Nuclear age, leading to a cold-war between USA and USSR. The resulting paranoia, particularly in the USA, must have led to fear of anything in the sky which couldn't be immediately explained. This would have worsened as mobile phone usage increased, and progressively more and more people became the owners of low-quality cameras. Also, the use of drones, both military and private, became more frequent. Nevertheless, not everything can be explained away as sightings of Venus, night-time drones, and experimental aircraft.

ii.   USA

There is a big difference between USA experiences and those of USSR/Russia as reported in the next section. Since the fall of Communism, official USSR government records have been made available to UAP researchers. Officially in the USA, UAPs do not exist and if anyone made a nuisance of themselves, they were ridiculed or threatened by government thugs – oops, sorry – Men in Black. Which is the totalitarian state here? It was only when some irrefutable USA navy video tapes[62] were accidentally released that they had to admit that some UAPs were actually real.

During the war Germany, who had sent expeditions to Antarctica previously, sent a further expedition there in 1938-9[63] despite all the other demands on its resources. It is suggested that this was connected to the crash-landed UAP in Germany in 1936. It was thought that there was an "Other" base there. Straight after the war, when it is possible that the USA could have acquired both the German UAP and the Italian UAP, the USA launched Operation High Jump[64] under the command of Admiral Byrd, to go to Antarctica. It is reported that Admiral Bird suddenly abandoned his expedition and hurried back the Washington after a confrontation with "Others".

When a massive NATO war-games exercise known as Operation Mainbrace[65] convened in the North Sea in 1952, it brought together 80,000 military personnel, 1,000 planes and 200 ships from nine countries. There were multiple sightings of UAPs which were documented by pilots and naval officers and appeared on radar.

From the initial detonation of a nuclear weapon on July 16 1945 onwards, there have been multiple reports of UAPs at test sites, nuclear facilities and ships carrying nuclear weapons. The most famous are possibly the Rendlesham Forest[66] Incident at an USA air base in England on 25 December 1980, and the USS Nimitz[67] film of 2004, where USA aircraft saw and videoed UAPs. When these videos came into public awareness, the USA government then acknowledged that they were genuine, but claimed that they didn't know what the objects were.

Back in the summer of 1952, a fleet of UAPs is said to have buzzed Washington in the USA. The first incident occurred on 12 September when fireballs were sighted over a wide area of the eastern USA. It is claimed that an UAP impacted the earth at Flatwoods[68], West Virginia (300km from Washington), with reports of extensive air-to-air combat[69]. Sadly, an USA fighter pilot is recorded as dying that night.

Then, before midnight on July 20th, an air-traffic controller at what was then Washington National Airport reported seven unexplained radar blips[70], including over the White House and Capitol. Other controllers—at National and at Andrews Air Force Base—and even a pilot in the air, also reported strange objects. One controller at National, according to a 2002 Washington Post account, looked out of the tower and saw a "bright light hovering in the sky" that resembled "a saucer" and then took off at "incredible speed". Fighter jets were launched to intercept them. On the 22nd in Silver Spring, Maryland (10km from Washington), a piece of debris was discovered which it was claimed had been shot from an UAP. The next weekend, the 29th, the odd radar blips were back[71]—this time, air-traffic controllers counted a dozen. The government tried to claim they were all the result of a temperature inversion disrupting the radar. This might convince a layman but was an insult to the expertise of the air-traffic controllers themselves.

On Feb. 20, 1954 President Dwight Eisenhower is claimed[72] to have interrupted his vacation in Palm Springs, Calif., to make a secret nocturnal trip to Edwards Air Force base, where he met with two ETs with white hair, pale blue eyes and colourless lips. These ETs are nicknamed "Nordics" in UAP circles because they resemble Scandinavian humans. They are said to come from the Pleiades, and are slightly taller than humans. The Nordics[73] offered to share their superior technology and their spiritual wisdom with Ike if he would agree to eliminate America's nuclear weapons. Ike declined the ETs' offer, because he did not want to give up the nukes unilaterally.

Sometime later in 1954, Ike reached a deal with another race of ETs, known as the "Grays" [74], allowing a permanent base in the USA, and permitting them to capture earthling cattle and humans for medical experiments, provided that they returned the humans safely home. "The Grays" is a convenient way to describe two gray-skinned species who often work together. These are imaginatively called the Tall Grays and Small Grays. They both supposedly come from the southern constellation Reticulum, and the Small Grays generally take the lead. The Tall Grays are thought to be a less-evolved version of the species. The Small Grays are the most observed "Others" on our planet.

The Small Grays are short, at about 1.0m - 1.3m tall, with a spindly body and legs and a large head with very big eyes. The Tall Grays are typically human height but with a larger head and eyes.

There appear to have been a number of UAP crashes in the USA since 1945[75]. Some of the early ones were The Trinity Incident[76] 1945, Roswell[77] 1947, and Kingman[78] 1953. In one of these, it is reported that an "Other" called J Rod[79], was found alive, and it contributed significantly to the USA's attempts at reverse-engineering flying saucers before it died some years later. Bob Lazar[80], a whistle blower in 1989, said that the USA held nine flying saucers which they had salvaged from crashes, and which they were reverse engineering.

What is difficult to understand is how a civilisation with technical ability to fly across the galaxy, can produce such terrible flying saucers that they keep falling out of the sky. This makes one wonder who was building these later versions – were some of these USA prototypes similar to those that had been seen by Jack Pickett[81] in a salvage yard at MacDill Air Force Base in 1967?

In December 1965, thousands saw an UAP leave a streak of fire across the sky, change direction and crash in woods near Kecksburg[82], Pennsylvania. Witnesses there described the machine as roughly acorn-shaped. They described how armed military turned up and took it away. Much later, it was claimed that the UAP looked like Die Glocke (The Bell), an UAP which was developed in Nazi Germany in 1945. The question is, of course, which was the chicken, and which was the egg?

Many people have claimed to have been abducted by "Others". They describe being subjected to medical tests, sometimes interfered with sexually, and sometimes finding some form of implant in their body afterwards. It is not known whether all abductees are returned.

There are many reports of animal mutilations, with horses and cattle being killed, drained of their blood and reproductive and facial organs removed[83]. These mutilations are often correlated with the appearance of UAPs and, sometimes, the dead animal appears to have been dropped from a height afterwards.

There have been multiple reports of large triangular UAPs the size of football fields, in the Hudson Valley[84] in the USA in the 1980s, in the so-called Belgian Wave[85] in the early 1990s, in Bakewell in the UK in 1993, Russia in 1989-1990, in the Phoenix Lights[86] incident in 1997, the Southern Illinois Incident in 2000, the Tinley Park Lights in Illinois in 2004-6, and the mass sighting at Stephensville[87], Texas in 2008.

There is believed to be a sprawling underground complex located beneath Archuleta Mesa on the Jicarilla Apache Reservation near Dulce[88], New Mexico. Dulce Base is said to contain at least eight levels that extend up to two miles below the surface. According to reports, the bottom three are staffed by as many as 18,000 Grays, as well as members of other species such as Reptoids – the "Others" base as agreed by Eisenhower? There are stories of a gunfight there in 2021, although no-one seems to know why, nor the outcome. It is also claimed there are tunnels for high-speed shuttle trains to connect Dulce to other similar sites across the western USA and that there are landing facilities for aircraft using UAP technology.

In 1964, a UAP was filmed during a US missile test[89] firing, attacking and destroying it over the Pacific Ocean. On March 27

1967, a UAP flew over the Malmstrom[90] Air Force Base in Montana, which housed ICBM silos, and this was followed by ten nuclear weapons going off-line for a short period. This may have been the Grays acting on their hidden agenda to stop all military nuclear activities.

On 17th November 1986, Japanese Airlines flight 1628[91], a Boeing 747-200F cargo aircraft, was en route from Paris to a layover in Anchorage, Alaska on its way to Tokyo. Pilot Kenju Terauchi, reported a seemingly inexplicable encounter in the skies as they approached the airport in Anchorage. He described a massive space craft, although no radars were able to confirm this. Over the next few months, further similar encounters were reported in the same area.

The Santa Catalina Channel[92] separates mainland California from the island of Catalina. This particular stretch of water is as deep as Mount Everest is high and UAPs have been seen both entering the water and emerging from its murky depths. But on June 14th 1992, there were reportedly hundreds of UAP's. Independent witnesses claim that there were in excess of two hundred UAPs that emerged from the water, hovering momentarily before accelerating off into the sky at blistering speeds, all the while in complete silence. There were many witnesses, some of which were as far away as Malibu and many filed reports with their respective local police forces,

The UK's tests of their trident missiles[93] in 2016 and 2025 which were conducted at a range in the USA, both failed soon after launch, leaving their test in 2012 as their most recent successful launch.

On 12th February 2023, an octagonal object with strings hanging from it was detected over northern Montana, Wisconsin, and the

Upper Peninsula of Michigan at 20,000 feet (6,096 m). Airspace was temporarily closed in the Lake Huron area, where the object was shot down by the US Air Force and National Guard, falling into Canadian waters[94]. This is reported to have caused a stand-off between USA and Canadian troops, over ownership but, as the USA got there first, possession proved to be nine tenths of the law. It was probably a Chinese spy balloon.

It is reported that many USA astronauts have seen UAPs following their capsules[95][96], but have been told not to talk about them. They can switch to a secure radio channel to report them, although some have been heard to refer to "bogies" outside.

iii.   Canada

As Canada has vast areas off wilderness, the level of UAP reporting is comparatively low, with most emanating from near the USA border, where the main centres of population are.

Labrador and Newfoundland account for a number of sightings, including from military and commercial aircraft. The most significant occurred on 1st February 1967 at Shag Harbour[97]. An aircraft was seen to crash into the harbour, and local fishing boats went out to look for survivors or wreckage. They could find nothing but an orange scum. Police, who had also observed the crash, then confirmed that there was no aircraft missing, and it was realised that they had a crashed UAP on their hands.

The next day, naval divers were sent down, and it was claimed that they had found nothing. It was realised that the UAP still had underwater manoeuvrability, and that it had crept out of the harbour and into an adjoining inlet, where it was discovered by the Canadian

Navy. However, it was not alone. A second UAP had arrived, and the crews were busy repairing the first. The naval vessels did not interfere, and eventually moved away in response to the detection of a USSR submarine many miles away. The two UAPs later rose out of the water and flew away to the north.

It is possibly not associated with the incident, but a lighthouse keeper nearby discovered a metallic object on the foreshore a few days later. He was not willing to say what was inside it, but he was ordered to hand it over to an USA officer who flew in specially to collect it.

iv. USSR/Russia

The main sources of reports of UAPs and Others are official government papers which were obtained after the fall of communism in 1989. The information here differs from that available from the USA in that, for long periods during the rule of communism and, after its fall, UAPs were officially recognised and the public and military were encouraged to report them. There was also an official committee to assess reports.

Russia has had its fair share of UAP sightings over the centuries, both before and after the Tunguska incident. It also had multiple sightings of Foo Fighters[98], in the skies and over the principal battlefields.

For many years, it was USSR policy to sent fighters to attack UAPs. However, they had already lost several aircraft before, in the 1960s, there was a full battle with UAPs[99] near their border with Iran and Afghanistan, where the USSR lost six fighters and their 12 aircrew. All fighter engagements with UAPs were banned in 1965.

There were multiple sightings over the Kola Peninsular[100] in the early 1980s, and there were reports of UAP crashes.

In October 4th 1982, UAPs appeared over a USSR ICBM base in Usove[101] in the Ukraine, performed astonishing manoeuvres in front of stunned eyewitnesses and then somehow took control of the launch system. The missiles which were aimed at the US suddenly started to initiate their launch procedures. Launch control codes were somehow entered, and the base was unable to stop what could have initiated World War 3. Then the UAPs disappeared just as suddenly, and the launch - control system shut down. This may have been the Grays giving a message to the USSR.

In 1983, in Ordzenikidze[102], east of the Black Sea, a cone-shaped UAP was shot down by ground-based weapons. The wreck was quickly spirited away by the Russians.

At Dalnegorsk[103] in East USSR, on 29th April 1986, an UAP was reported to have crashed into a hill (known as Hill 611). There were multiple UAP sightings over that hill thereafter, and the crash-site was somehow dangerously active for 3 years, with adverse effects to humans. Materials recovered were described as complex and not yet available on earth.

On April 26th 1986, the Chernobyl[104] disaster occurred. Two technicians, who were dashing to the site after the first alarm, had to turn back when they measured the very high radiation. As they retreated, they saw a spherical UAP appear, which fired two red beams at the reactor for about 3 minutes, before departing. The technicians reported that the radiation levels fell from 3,000 milliroentgens/hour, to 800 milliroentgens/hour in that 3 minute period.

Although there had been occasional reports of UAPs buzzing USSR nuclear facilities before this event, their frequency increased significantly afterwards.

Russian Cosmonauts reported that, in May 1981, the space-station Salyut-6[105] was approached by an unidentified space-ship to within 300 feet, and they could see 3 brown-skinned beings with slanted blue eyes. They communicated by holding up pictures to the windows, but didn't really get anywhere. The "Others" came out of their ship without any form of protective suit or breathing apparatus. It is thought therefore that these might have been some form of AI. The whole episode was filmed, but this evidence is not available.

Cosmonauts have also reported seeing angel-like entities[106] made of plasma, outside their space-station. At first, it was thought that this was caused by isolation or claustrophobia, but their replacement crew reported exactly the same phenomenon.

In 1989, Russia sent two probes to the Mars moon, Phobos[107]. Unfortunately one never made it, and the second was lost soon after arrival, but not before it had photographed an UAP which was a cylinder 15 miles long.

UAP sightings have continued and, since the outbreak of the Russian-Ukrainian war, there are regular sightings over the battlefields.

v.   United Kingdom

UAP sightings were not recorded at a national level in the UK until the 1950s[108], when 13 were recorded in that decade, including one at RAF Bentwaters, later to become famous as the site of the Rendlesham Forest Incident of 1980.

On 4th February 1977, it is claimed that a silver UAP landed on a field at the back of Broadhaven Primary School[109] in Wales, where it was seen by 14 pupils. The next day, their head teacher asked these to draw what they had seen, and the drawings were all remarkably similar. The same thing occurred on 16th February in an Anglesey Primary School, where 9 pupils drew very similar versions of what they had seen. However, this second sighting is considered less reliable, because it could so easily be a copycat report.

On 9th November 1978 a forestry worker claimed that an UAP had attempted to abduct him[110]. The police investigated and confirmed that the marks on his body, his clothing and the site where it occurred, all fitted in with his story, but could do no more. This occurred in what is now called the Falkirk Triangle[111], because of the large number of sightings recorded there. It is claimed that there are about 300 sightings there each year. This begs the question why? There may be an "Other" base nearby, and it is not far from Faslane, which is the home base of the UK nuclear armed submarines.

On 26th December 1980, Just after midnight, eyewitnesses and radar screens at RAF Bentwaters, a USA air base reputed to store nuclear weapons, followed an unidentified object as it vanished into Rendlesham Forest[112]. Soldiers dispatched to the landing - site encountered a small luminous triangular - shaped craft, ten feet across and eight feet high, with three legs. The UAP then retracted the legs and easily manoeuvred its way out through the trees. The soldiers chased it into a field, where it abruptly shot upward, shining brilliant lights down on them. At that moment the witnesses lost consciousness. When they came to, they were back in the forest.

Other troops sent to rescue them found tripod landing marks where the object apparently had rested.

The following evening, after observers reported strange lights, the deputy base commander, Lt. Col. Charles Halt, led a larger party into Rendlesham Forest. There, Halt measured abnormal amounts of radiation at the original landing site. Another, smaller group, off on a separate trek through the forest, spotted a dancing red light inside an eerily pulsating "fog." They alerted Halt's group, who suddenly saw the light heading toward them, spewing forth a rainbow waterfall of colours. Meanwhile, the second group now watched a glowing domed object in which they could see the shadows of figures moving about. During the next hour both groups observed these and other darting lights.

There were large-scale investigations into the incident by both the USA and UK, with claims of USA agents using drugs to interrogate personnel. It has been claimed that the whole thing was a psi-ops test, but it seems unlikely that the USA would try this on the guards to an important military facility, risking their sanity and hence the security of their weapons, when they could do the same thing in home territory.

In 1997, the UK government received over 450 UAP sighting reports[113]. The UK public had clearly found someone to tell. This compares with only 9 in the whole 1970s decade. Of these sightings, there were about 30 large triangular craft, and the remainder were mainly bright lights or flying saucers, with a few cigar-shaped craft.

On 20th July 2008, a police helicopter had to take evasive action to avoid a UAP flying over the Bristol Channel[114], near the base at RAF

St Athens. There were also a number of sightings reported over South Wales that day.

On 22nd February 2016, residents of Pentyrch, a village in Wales, NW of Cardiff, reported a large pyramidal UAP[115], which hovered at a low level and discharged two smaller objects. Over the previous week, there had been reports of RAF Sentry aircraft regularly sweeping the area but now, a whole fleet of UK military aircraft appeared, and the UAP moved away. Two large explosions were reported. A Freedom of Information request for more details was refused on a variety of grounds. It is apparent that this UAP had been expected, and wasn't welcome.

Over the period 2020 – 2024, UAP researchers gathered information from the police using Freedom of Information[116] enquiries and found that there were about 500 claimed sightings each year. The problem clearly has not gone away.

vi.   Mainland Europe

In France, on October 16th 1954, a Dr Robert[117] saw 4 UAPs, one of which landed in front of his car, and a small "Other" came out. All went dark for some time until the UAP flew off.

On October 27th 1954, an UAP passed over a stadium soccer match in Florence[118], Italy. It was witnessed by over 11,000 spectators as well as people outside the stadium.

In October 1958, a motorcyclist in Catalonia[119], Spain, stopped to assist a crashed plane. Instead it turned out to be a landed UAP with two small "Others" outside collecting samples from the local terrain, with one more inside. The witness watched for 15 minutes until they flew off.

In Belgium[120] on September 22nd 1965, several witnesses, including Antwerp Airport employees saw a glowing sphere-shaped UAP flying at an estimated speed of 3,000 kph.

Over Donaghadee in Northern Ireland over a dozen people sighted several egg-shaped UAPs travelling at a phenomenal speed on 4th November 1975.

From November 1989 to March 1990 there was a wave of triangular UAPs over Belgium[121]. Many of the reports related a large object flying at low altitude. Some reports also stated that the craft was of a flat, triangular shape, with lights underneath.

The Belgian Wave peaked with the events of the night of March 30th, 1990. On that night, one unknown object was tracked on radar, and two Belgian Air Force F-16s were sent to investigate, with neither pilot reporting seeing the object. No reports were received from the public on that date but, over the next 2 weeks, reports from 143 people who claimed to have witnessed the object were received. Over the ensuing months, many others claimed to have witnessed these events as well. Following the incident, the Belgian Air Force released a report detailing the events of that night.

On 2nd September 1990 in Greece, there were reports of a UAP crash near the village of Megaplatanos[122] close to Atalanti. Around 3 am, shepherds and a small group of villagers observed 6 UAPs approaching from the north. One of the UAPs looked unstable. Bizarre lights emitted from the ships which did not make any noise. The troubled UAP lost altitude and crashed 500 meters or 1/3 of a mile from them. They did not hear any noise although the wooded area caught fire. The remaining UAPs stopped over the accident while two landed near the destroyed UAP. The fire was instantly put out and during the entire night, there was unusual traffic from the ground to the hovering UAPs, light spots went up and down collecting debris until sunrise. The UAPs disappeared right before sunrise. The entire village saw the event. The ground featured an oval-shaped burn with a cut pine tree in the centre. There were also very small metallic pieces and wires around the crash site. The Hellenic Air Force arrived hours later and informed the village that it may have been a Soviet satellite that crashed or a plane. The Hellenic Air Force took some of the pieces.

On November 5th 1990, a triangular UAP[123] was spotted by air traffic controllers over Paris, Bischwilleer and Nantes. The UAP was also sighted by three aircraft crews in Italy.

In 2009 on 9th January at 1 AM, Adam Maksymów, of the village Jarnołtówek[124] near Prudnik in Poland, went outside to charge his car battery. He was interrupted by a noise which he likened to rockets blasting off. Then he heard a buzzing sound which he compared to that of a swarm of bees. He saw a blinding light, and a huge saucer with a triangular glowing blue beam on its underbelly rose above the ground and took off into the night sky at an impossible speed. Other people of Jarnołtówek also reported seeing the object.

I have only listed here a tiny fraction of the total list of UAP sightings across Europe. Overall there are thousands.

vii. Central America and the Caribbean

To the west of the main island of Cuba lie the waters of Guantanamo Bay[125]. Apart from its more notorious current reputation, there were also sightings of UAPs there in the 1960s.

Puerto Rico[126] is described as a massive UAP hot-spot, with the possibility of an UAP base beneath the lake Laguna Cartagena[127] and a second, underwater, to the north in the Milwaukee Deep. There have been multiple reports of UAP sightings, a child-abduction and sightings of a number of Chupacabras[128], or Goatsuckers[129].

Could it be that these Chupacabras are the actual "Others" here? They don't appear to harm humans, and it may be that they have a need to ingest blood to survive. This could be part of the reason that there are world-wide reports of cattle, sheep and horses being found drained of their blood. Are we here on earth helping to keep alive a species that needs blood to eat, but has the conscience not to attack intelligent species?

There have been many sightings of UAPs[130] apparently flying into Popocatepetl near Mexico City. This is an active volcano, but it is not the only one associated with UAPs. They could be openings to Portals or Gateways, bases for fire-loving ETs, or homes of ITs.

On August 25th 1974, north of Presidio[131], it is claimed that an UAP travelling at 2,000 miles per hour collided with a small plane. The flaming wreckage of both aircraft fell to the ground and the small plane was pretty well destroyed in the crash, but the UFO was intact, When the Mexican Army arrived, they found pieces of the plane and a silver disc approximately 17 feet across and six feet tall. Twenty-four Mexican soldiers surrounded the craft, amazed at their find, loaded it onto a flatbed truck and headed back to their base at Ojinaga. An hour later, all 24 soldiers were dead. The truck veered off the road and rolled to a stop. The US was monitoring all this, listening in on radio communication between the soldiers on the ground and their headquarters. They flew a reconnaissance plane over the site of the truck and bodies. A flight investigation team was assembled and four helicopters took them from Fort Bliss in El Paso to an area about 20 miles south of Candelaria, Texas. The U.S. team, wearing biohazard gear, took it away. They then set off explosives to cleanse the area of contamination. The UAP was taken to a secret government base in the Davis Mountains, and then transferred to another military compound, probably Wright Patterson Air Force Base near Dayton, Ohio.

On July 11th 1991, there was a total eclipse of the sun over Mexico, and thousands reported seeing UAPs[132]. There were many films of the event. Since then, there have been thousands of sightings throughout Mexico. There is no clear pattern, although the majority have occurred in and around Mexico City. The witnesses range from pilots to doctors to bus drivers and even school children.

On 5th March 2004, a Mexican military aircraft[133] filmed a line of 11 UAPs flying near it, using a state-of-the-art infra - red camera. At one stage, the objects surrounded the aircraft. The Mexican government confirmed that it had released the video.

viii.　South America

There are claimed to have been sightings of large spacecraft hovering over generating plants and reservoirs there, causing power blackouts on a regular basis. Some tribes complain that they are stealing their water.

In Argentina, on 10th May 1950, it is claimed that a 32 feet diameter UAP crashed[134], killing the three "Other" occupants who were about 4 feet tall. The witness was an aeronautical engineer, who entered the craft, then left to get help. When they returned, there were two further UAPs overhead, and the crashed one had been reduced to a pile of ashes. Afterwards, he suffered from a fever for some weeks, and had blisters on the parts of his face which were not protected by the sun-glasses he was wearing.

In Venezuela[135], on 28th November 1954, it is claimed that two hairy dwarf-like creatures and an 8-10 feet luminous sphere stood in the way of two truck drivers. One driver attempted to capture one of the dwarves, and lost the fight, fortunately with no major injury. The sphere flew off. On 10th December[136] there was a further sighting claimed, where dwarves again attempted to capture a young man, who managed to hit one with his unloaded shotgun, which shattered. The sphere flew off again. A similar incident is reported for the 16th December.

Back in Argentina[137] on 20th August 1957, an Air Force guard claimed to have been approached by a disc-shaped UAP, from which came a voice claiming they were interplanetary beings who were setting up a base in Salta, to liaise with humans.

That year there were numerous reports of UAPs in Brazil, including one reported by Professor Guimares[138] who was approached by an UAP, from which two tall beings descended. He was taken on a short flight.

On 5th July 1965 and thereafter, in the Valle of Loretani[139], in Argentina, there have been repeated sightings of UAPs about 10 –15 metres in diameter, seen by dozens, including the head of the Argentine's UAP investigations, and the frequency has led to the suspicion that there is a base in the area.

In Peru in 1968, there were many UAP sightings, leading to suggestions that they were using areas around Lake Yanacocha, Lake Titicaca[140], and various other lakes in the Cordillera Blanca.

On January 13th 1996, the Brazilian Air Force is alleged to have shot down an UAP which crashed six miles from Varginha, a medium-sized town in south-eastern Brazil[141]. Seven days later, two sisters aged 14 and 16, and a 21-year-old friend spotted a tiny, frightened alien with big red eyes, crouching by a wall. They ran screaming back to their mother. The Brazilian police and military captured at least two aliens, one of which scratched an officer, infecting and ultimately killing him, before dying along with its extraterrestrial comrades. The US Air Force confiscated the alien bodies and took them to an unknown location. A vast cover-up by the Brazilian military, enforced with death threats, lasted for 26 years.

The above examples are just a few of the many alleged sightings recorded in South America in this period. With the growth of air travel on the continent, there has also been a commensurate growth of claimed UAP sightings in the air, including the sighting on 5th October 1996 in Brazil[142], of a massive UAP, cylindrical in shape, and 300 – 400 feet long.

In Argentina on the night of 5th September 2023, in the Espora Air Base, near Bahía Blanca[143], four UAPs flew over the base and were seen by military personnel, who responded with gunfire. The objects were black and triangular in shape. One object fired with a type of laser that injured two or three soldiers. The Base denied any incident and claimed that the videos and audios that circulated that night were edited to create fake news.

ix.   Japan and the Dragon's Triangle

The Dragon's Triangle (Devil's Triangle) is loosely defined as an area south of Tokyo, to the Philippines and the Marianas Islands. However, in his book on the subject, Charles Berlitz[144] even looks at incidents in Tasmania, south of Australia and also in New Zealand! Great care needs to be exercised in using his findings.

The Dragon's Triangle has a reputation at least equal to the Bermuda Triangle, with similar fuzziness about what lies within its limits, and what is outside. Many of the incidents recorded there may not involve UAPs, as it has notorious waves and underwater volcanoes, but many believe that there is an underwater UAP base there.

On February 21st 1957, the citizens of Yokohama City[145] watched as fleets of UAPs flew over them in V formations. The whole affair lasted seven minutes, and they were totally silent.

When sailing back to Tokyo from the south[146] on April 19th 1957, a fishing boat encountered two metallic very silvery UAPs descending from the sky, which dived into the water.

On March 21st 1965[147], an UAP followed a TOA airlines plane for 55 miles, causing some electronic systems to fail. The pilot was able to use his radio to tell traffic controllers what was happening and a second aircraft then claimed that the same thing was happening to it.

In 1975 on 23rd February[148], two seven year old boys watched a bright silvery UAP, shaped like a domed disc, land in a vineyard in Kofu. In it was a humanoid creature, dark skinned with pointed ears, and a wrinkly face with no other features apart from fangs. It spoke unintelligibly and the boys ran off. Their parents saw orange light in the vineyard as the UAP flew off.

When sailing in the Devil's triangle on 17th April 1981, a freighter, the Taki Kyoto Maru[149] suddenly lurched as if it had been struck by something. A 50 feet diameter flying saucer rose out of the sea, and circled the ship for about 15 minutes. It then dived back into the sea, causing waves which almost capsized the ship. The captain then found that the ship's clocks had lost those 15 minutes.

Between 1946 and 1986, there have been a total of about 25 large surface ship disappearances[150] in the Dragon's Triangle. In that same period a total of 13 soviet submarines were lost in the Japanese area, together with any nuclear weapons which they were carrying. However, only 2 of these submarines were in the Dragon's Triangle. The others were in the Sea of Japan or the South China Sea.

Close by this arbitrary triangle is the island of Pohnpei in Micronesia. This contains the remarkable ruins of Nan Madol,[151] and nearby are claimed to be two legendary submerged prehistoric cities known as Kahnihmw Namkhet and Kahnihmweiso. There is little archaeology to support this claim, but it could be that this was the location of an "Other" base.

In the mountainous region of Fukushima Prefecture[152] in Japan, residents report frequent sightings of luminous objects near Senganmori Mountain, and of "Others" with wings and hawk-like features. These Avians sound similar to Anzu of ancient Sumeria or Horus of Egypt.

x.   Australia & New Zealand

On January 19, 1966, a calm sunny day, a banana farmer named George Pedley was driving a tractor near Horseshoe Lagoon[153] near Tully, in tropical far north Queensland, Australia. When he was about 25 yards from the lagoon, he heard a loud hissing sound above the noise of the tractor. Suddenly, an object rose out of the swamp. When he glanced at it, it was already 30 feet above the ground, and at about tree-top level. It was a large, gray, saucer-shaped object, convex on the top and bottom and measured some 25 feet across and 9 feet high. It rose another 30 feet, spinning very fast, then it made a shallow dive and took off with tremendous speed. Climbing at an angle of 45 degrees it disappeared within seconds in a south-westerly direction.

When he came to the spot from which the object had risen, there in the lagoon was a large circular area that was clear of reeds and in which the water was rotating slowly. A few hours later, at about noon, he returned to the lagoon for a second look. The scene had changed, because now the circular area was covered by a floating mass of green reeds that were distributed in a clockwise radial pattern. The circular mass of reeds was about 30 feet in diameter.

Oddly, the outside edges of the mass of reeds angled down, similar to the shape of a saucer placed face down. He reported his experience to the Tully police that evening, and they in turn reported it to the RAAF after making a trip to the site the next day, January 20th. During the course of the investigations, as many as five other "nests", all smaller than the original, were discovered. In some of these, the reeds were rotated in a counter-clockwise direction and a couple of them showed signs of burning in the center of the nest.

On 6th April 1966, students and a teacher from Westall High School[154] (now Westall Secondary College) reported seeing a flying object. It was described as round with a domed top, and white, gray, or silver in colour. According to the students, the object descended behind a row of trees and into the Grange, an open area south of the school. Some accounts describe the object as being pursued by five unidentified aircraft. Shaun Matthews was on vacation at the Grange and reported seeing an object with a slight purple hue and about twice the size of a family car.

Some witnesses reported seeing the object take off after landing, and some reported seeing it hover rather than land. When students walked to the Grange after the sighting, some reported a landing site, but the details varied between reports. Students variously described a circle of grass as burnt, "boiled" or pressed down. One student interviewed by a local newspaper described a vague circular area flattened by the wind. Students also reported varying numbers of circles from one to three.

The Kaikōura lights[155] is a name given by the New Zealand media to a series of UAP sightings that occurred fist on 21st December 1978, over the skies above the Kaikōura mountain ranges in the northeast of New Zealand's South Island. The first sightings were made when the crew of a Safe Air Ltd cargo aircraft began observing a series of strange lights around them, which tracked along with their aircraft for several minutes before disappearing and then reappearing elsewhere. The UAP was very large and had five white flashing lights that were visible on the craft. The pilots described some of the lights to be the size of a house and others small but flashing brilliantly. These objects appeared on the air traffic controller radar in Wellington and also on the aircraft's on-board radar.

On 30 December 1978, a television crew from Australia recorded background film for a network show of interviews about the sightings. For many minutes at a time on the flight to Christchurch, unidentified lights were observed by five people on the flight deck, were tracked by Wellington Air Traffic Controllers, and filmed in colour by the television crew. One object reportedly followed the aircraft almost until landing. The cargo plane then took off again with the television crew still on board, heading for Blenheim. When the aircraft reached about 2000 feet, it encountered what appeared to be a large lighted orb which fell into station off the wing tip and tracked along with the cargo aircraft for almost quarter of an hour, while being filmed, watched, tracked on the aircraft radar and described on a tape recording made by the TV film crew.

They have appeared intermittently since the initial December 1978 sightings, with the most recent sighting being reported during 2015.

A Darlington[156], Perth, man says he's captured pictures of hundreds of UAPs from the veranda of his home. It began when he was taking photos of clouds to test out a new camera and he noticed a "smudge" that, when enlarged and enhanced, "had some structure to it, suggesting it could be some sort of craft in the sky". He says that, since then, he has identified a dozen different UAPs including round, square and saucer-shaped craft, posting the photos to his website wispyclouds.net for extra-terrestrial buffs and sceptics to ponder.

He takes about 30 shots at a time. In 10-15 minutes he'll take 300 to 400 images. Then he connects his camera to the computer, zooms in and enhances any little thing he notes on the images. He gets UAPs in anywhere from 2 per cent to 20 per cent of shots. Some of them appear to have transparent canopies and in some shots it looks like there could be occupants inside. There is always doubt, but UAP stands for unidentified anomalous phenomenon and as far as he's concerned these aren't identified. It's possible some are man-made, but he doesn't think they all are.

Pine Gap[157], near Alice Springs, was originally set up as a USA eavesdropping station to listen in to China. However, it has been much enlarged with great warrens excavated beneath it, and it is now thought that it is acting as a liaison base for "Others". There have been a number of reported phenomena.

One of the strangest incidents occurred in 1973[158]. A cartographer, working for the Australian government, was parked near to the Pine Gap base. It was late – just after midnight. Out of nowhere, a strange but intense "vertical shaft of blue light" came from the confines of the base. Giving in to his curiosity he drove his vehicle closer. He was shocked to see a disc-shaped craft hovering around a thousand feet in the air over the base. He raised his binoculars to his eyes. As he did, another blast of cold, blue light emerged. It was coming from the center of the craft and heading down to the domes of the base. After several moments it went off. Then, a similar laser-like light extended from somewhere on the ground to the disc above. This went on for over thirty minutes before the disc began to spin rapidly. It then shot off into the night sky at great speed.

At some time during 1984, an intense gold pillar of light[159] shot upwards from the middle of Pine Gap base grounds. It was several meters wide and appeared to be solid as it stretched upward into the night sky. Above them, and around the point of the light, a strange cloud appeared to be forming. As the light "pulsed" the cloud appeared to grow. Just as suddenly as it began, the light shut off. Before the five witnesses could get their bearings, however, the light appeared again, only this time coming down from above and hitting the grounds of the base. It was at this point that the witnesses saw for the first time five strange objects above the base. Four of these were in a "diamond formation", while behind them was a cylindrical object. The first four objects moved into north-east-south-west positions, while the cylinder-shaped object positioned itself in the middle of them. Then an intense solid light, this time blue, hit the ground from the objects above. Then strange cloud formations also appeared high up the beam. This continued for several minutes before the lights shut down and the five apparently alien crafts slowly rose upwards before vanishing in a flash. Although none of the witnesses knew what they had just seen, all agreed that the rumors of established contact with an extra-terrestrial race were apparently true.

At around 4:30 am in the early hours of 22nd December 1989, three men, while returning from an all-night hunting trip[160] near to the Pine Gap facility, suddenly noticed activity in the grounds of the base. A camouflaged door suddenly opened before them, revealing lights and movement behind it. From inside this hidden enclosure, a metallic, gray disc emerged. No sound accompanied its movements, and aside from their own breathing, all around them was silent. Suddenly, but still with little sound, the disc shot off at an amazing pace, certainly beyond anything any of the witnesses had ever seen. The door then calmly shut, hiding its presence once more in the process.

xi.   Worldwide

UAP appearances are in no way confined to the various areas I have detailed above. They are a worldwide phenomenon as are portals and gateways. Cattle mutilations occur all over the world, and not just of cattle. Both sheep and horses are subject to attacks as well. It would appear that ETs did not really need permission to take cattle in the USA. They would have done it anyway.

China has a long history of their larger mountains being considered "Other" bases, with UAPs being sighted in their vicinity. It has been suggested that the Chinese concept of a dragon comes from sightings of flame-spouting UAPs being miss-interpreted. China's first Emperor Qui Shi Huang is thought to have either been "Other" or closely assisted by "Others". There have been recent photographs of apparent UAPs over major airports[161], resulting in their temporary closure.

In Africa, the principle items of interest are the pyramids in Egypt and the Sphinx. However, there are many reports of lights in the sky and of circular UAPs of various sizes and a couple of reports of large triangular UAPs. There is at least one claim of an UAP landing in 1989, in the Kalahari Desert[162], where cattle herders described strange very tall humanoid "Others" exiting. They wisely fled. Nearby the same year, it is claimed that a UAP crashed[163], with a South Africa military cover up.

Worldwide, wherever there is still a tribal sense of identity, as in Africa, part of their oral tradition is that "Others" came to teach them. In places, rock art preserves a record of this, as well as carvings.

xii. Summary

Having looked at the vast number of sightings that have been recorded around the world, I am left in no doubt that UAPs are real. Many of the sightings are of lights in the sky, rather than of identifiable spacecraft, but these lights are simply "Other" reconnaissance drones, of vastly superior abilities to those we can make for ourselves. There are enough sightings of larger UAPs to justify my conclusion. Beyond that, the attitude of governments towards them, makes it clear that they know all about them, and are either content to leave them to their own devices, or are too frightened of them to do anything.

d) WHEN THEY MAKE CONTACT

i. Contact with the Authorities

Now that I have come to the conclusion that UAPs are real, I also have to conclude that they have been in regular contact with many governments since at least World War II. It is not my place to justify their reasons for withholding this information.

ii. Contact with Real People

In June 1920, Albert Coe[164] was canoeing in Ontario when he rescued a stranger from a ravine. He had to carry him to his craft which turned out to be a 20 feet diameter flying saucer. He watched it take off, and saw him afterwards 10-12 times a year until the 1970s. The "Other", named Zret, had crash-landed on Mars with 3700 "Others," eventually colonising Earth in Atlantis, the Cuzco Valley in the Andes, Lemuria (near the Marshall Islands), Tibet and Lebanon. Their main current concern was Earth's potential to develop atomic weapons. He looked fully human.

In May 1940, a tall humanoid with white hair descended from a flying saucer at Helena[165], Montana in the US, and asked for assistance in getting water from a nearby stream. They said that they were there to monitor the development of humans, but were not allowed to interfere. They were probably Nordics.

On 4th July 1949, Daniel Fry[166], a rocket test technician at White Sands in the USA, saw an UAP land and, as he approached, a voice warned him not to touch as it was dangerous. The voice turned out to be that of an "Other" who called himself Alan, although he never saw him. He said his purpose was to see how adaptable humans were to new ideas. Daniel was given basic descriptions of how the UAP's mechanism worked. He was also told how they had developed a previous earth civilisation on Atlantis, which had been destroyed by war, and the survivors had fled to Tibet.

On 11th April 1952, near Nimes in France, Rose C[167] was woken by her dogs and, going outside, met 4 "Others", three being very tall. The leader explained that they had established earth thousands of years ago, as a penal colony. About 6400 BCE, there was a man-made cataclysm on earth, destroying their civilisation. Their job now was to take vegetation and soil samples following our use of atomic bombs.

On 20th November 1952, George Adamski[168] claimed to have met the occupant of a cigar-shaped UAP in the Californian Desert. Called Orthon, he was long haired and about 5' 6" tall, and said he was concerned about radiation from nuclear bombs. Adamski was invited to board the UAP but declined and Orthon flew off. Adamski then claimed that Orthon returned, and started trading on his sighting, and claimed that he had gone aboard the UAP. Over time, his claims became more and more extravagant, until eventually his most ardent supporters gave up on him. His secretary left in the

1960s, citing his oversize ego as the reason why the "Other" gave up on him.

When he was 10 years old, in 1932, Howard Menger[169] first encountered an "Other", a beautiful tall young woman with golden hair – probably a Nordic. He had further encounters with male and female Nordics regularly for the next 65 years. They told him of the coming of atomic bombs. They asked him to acquire earthly clothing and fresh food (although they weren't too keen on it). They took him on trips to the moon, visiting their base there. He was taken to their earth base in the Blue Mountains of Virginia, where it is believed they were mining beryllium, zirconium and titanium. Although he enjoyed the notoriety when he first spoke about his experiences, he quickly came to regret it.

iii. What to believe?

The stories in the section above could all be fantasies dreamt up by attention seekers. The "Others" described here all seem benign, and it may be that you don't hear of malignant "Others", because no contactee survives them. Nevertheless, these are all described as having concerns about our nuclear weapons, and our polluting our world: and we have no evidence that our governments have the same worries.

Chapter 3

SO WHAT ARE THESE "OTHERS" ?

a) <u>INTRODUCTION</u>

It is probably far more likely that UAPs are "Other" in origin, although this does not mean that they are necessarily Extra Terrestrial. However, this still gives far too many choices:

- They are caused by gods or magicians
- They are from our future
- They are from an earlier human civilisation in hiding
- Apes weren't the only earthly creatures to evolve
- They come from parallel universes
- They are products of an "Other" culture
- They are human products of reverse engineering of "Other" technology

b) <u>THEY ARE CAUSED BY GODS OR MAGICIANS</u>

Most folklore talks about Gods. They generally descend to earth and do amazing things, often to the benefit of the local humans. Some stories imply that these Gods can get very angry, and even fight each other[170].

The distinction between a God and a Magician is a matter of personal belief. They can both do apparently impossible things. But, as Arthur C Clarke said "Any sufficiently advanced technology is indistinguishable from magic". I have always wondered why, if a being is as far technically advanced from me as I am from an ant, they should want us to build churches and worship them. I would be horrified if I found that ants were worshipping me.

So for the moment, as I am not clear what makes a being a God, I am simply going to define Gods as belonging to a superior civilisation.

c) THEY ARE FROM OUR FUTURE

There is little evidence to support this hypothesis, apart from some photographs of anachronisms, which could be simple miss-interpretations. The classic possibility of time paradoxes suggests that it would be unwise for our descendents to try to change us lest, in doing so, they change themselves. There is the possibility that they could be sight-seers from the future. There are unconfirmed reports of UAPs watching the first moon-landing.[171] As this was one of main scientific events of the twentieth century, I would love to have had the chance to watch!

d) THEY ARE FROM AN EARLIER HUMAN CIVILISATION, IN HIDING

Given that the time interval between the extinction of the dinosaurs, and the generally accepted evolution of the modern human some 20 thousand years ago, is some 66 million years, it is not beyond the realm of possibility that a fraction of modern humans actually started their final stage of evolution say 30 thousand years ago. This would give that fraction a 10 thousand year lead on most humans in developing civilisation. They would today appear very like the rest of humanity, and could co-mingle without detection.

In the extra 10,000 years of their civilisation's development, they could have fully explored the Solar System, and could be living out there, leaving the earth to us apart from the occasional hidden base. They could even have been the "Gods" who came back to educate early mankind.

Equally, as described in the previous chapter, they could have been wiped out at some stage. Several "Others" have told contactees this. What is missing from archaeological records is evidence of their using resources beyond stone. If they had developed a society similar to our present one, there would be mineral mines for us to find. They would have to have gone down a different technological route which did not involve the use of metals to any great extent.

e) APES WEREN'T THE ONLY EARTHLY CREATURES TO EVOLVE

Whilst humans were evolving in the 66 million years since the extinction of the dinosaurs, all the other creatures on earth were evolving too. To have avoided the dinosaur extinction event, they would probably have to have lived deep underground or deep underwater, may even have developed more than primitive mammals before the extinction event, and could have got a running start towards civilisation. They could be much more advanced than humans.

If they lived in very deep caves, or underwater beyond the continental shelves, it is quite understandable that we have not yet found any archaeological record of their existence.

These are what are termed Intra-Terrestrials (ITs).

f) THEY ARE PARALLEL UNIVERSE ENTITIES (PUEs)

Physicists, working in the field of String Theory, seem happy to contend with multiple universes, perhaps even an infinite number of them. It may be possible to move from one universe to another via Portals.

One proposal resulting from string theory is that, whenever for example there are two possible answers to a decision, the universe splits into two new universes, one for each possible outcome. As this is happening everywhere, all the time, we end up with an infinite number of possible universes. This poses a problem when we want to move between universes. In many cases, it would not be possible to tell one universe from many others, until the results of all the possible outcomes have worked through enough to show a significant difference. This might take years. Even if one developed a way to move from one universe to another by a Portal, it might be very difficult to find one where things were drastically different. How long ago would the change have to have happened before a universe had a significantly different technology from ours, let alone to appear physically different?

If we are the subject of incursions from a parallel universe, this would suggest that there might be far fewer such universes. The inhabitants of one of these would need to be able to create Portals, through which they could visit us.

If Portals between parallel universes could open of their own accord, this might account for some of the exotic creatures which are claimed to exist, but do not appear to be in sufficient numbers to constitute a breeding population. These cryptids might include Mermaids, Skinwalkers, Unicorns, Werewolves, Mothmen, Chupacabras and many more. Of course, they may simply be entities here for a vacation! It is difficult to imagine secret bases for Reptoids in the swamp areas they appear to enjoy, but they are reputed to be intelligent, if aggressive.

I have deliberately excluded Bigfoot[172], (including Yeti, Sasquatch, Yowie, etc) on the grounds that there have been so many sightings that, even without any definitive proof of their existence, it is credible that there could be sufficient numbers present for a breeding population to exist.

Small Portals are sufficient to permit the passage of the small UAPs generally described as globes of light or small metallic objects, which seem to be reconnaissance devices to see "what are these humans up to now?" Native Americans have shown that they are capable of opening small portals using rhythmic chanting and drumming. It is claimed that this was filmed in the TV series "Beyond Skinwalker Ranch". In the Indian sub-continent, the use of chanting appears to serve the same purpose, with mantras claimed to be very powerful.

Larger Portals would be necessary to permit the passing of larger UAPs, which could be described as means of transport for entities of approximately human proportions. This might generally require more energy than could be achieved by drumming and chanting, but could still require some sonic trigger.

If some parallel universes could be considered to be near each other, it might be possible for events in one universe to have an effect on another. This might account for reports of PUE visitors being concerned about our use of nuclear weapons, and their being particularly interested in nuclear tests, nuclear stockpiles, nuclear-armed rockets, aircraft and ships, and nuclear power plants. Whilst any race in any universe or galaxy may well be altruistically concerned that we shouldn't wipe ourselves out, they may well be much more concerned if we might wipe them out too.

Certainly, if we find signs of crossing points into our world, we don't have to assume they are wormholes, or that our visitors are from outer space.

g) THEY ARE PRODUCTS OF AN "OTHER" CULTURE

i. General

Firstly we need to remember that there are many possible different types of "Other" culture. As explored in the previous section, there is the possibility of a large number of cultures in parallel dimensions (PUEs), which could reach us by Portal. Secondly, if long-distance travel within our own dimension is possible, whether or not by Wormhole or Gateway, there could be an almost infinite number of planets where Extra -Terrestrial (ET) cultures might be much more technologically advanced than us. Thirdly and often ignored by Ancient Astronaut Theorists, there is the possibility of a limited number of Intra-Terrestrial races (ITs) living on this planet either deep underground or deep underwater, and which could again be more advanced than us. They would certainly be concerned by our using nuclear weapons.

It is very concerning to hear claims that Occam's Razor can be applied to prove that "Other" cultures, which have been visiting us for millennia, must be ET. The simplest solution should not be the one which requires the most complexity.

ii.   Intra-Terrestrials (ITs)

It is quite likely that this possibility has not intruded greatly on earthly consciousness. After all, we should have seen them, or their bodies, or archaeological evidence, shouldn't we? This is not necessarily so.

Let's begin by considering things from a historical perspective. The dinosaurs were wiped out approximately 66 million years BCE both on the land and in the sea. At that time there were small mammals which survived, and from which it presumed that humans and other mammals evolved. Presumably these small mammals would have to have lived underground to have avoided the extinction events. The skull of Lucy, thought to be an early hominid, was dated to 3.2.million years BCE. Between these dates, the last ice-age started about 34 million years BCE and continued until relatively recently at about 11 thousand years BCE.

66 million years is the time it took for humans to evolve, but who is to say that other species – ITs - did not evolve in the same period, perhaps more rapidly than us? There are apparently no fossil records to suggest this. However we have not yet searched for fossil records in ultra deep cave systems or the depths of the oceans.

Native American Hopi folklore speaks of Ant People[173] living deep underground, possibly somewhere in the Rocky Mountains. African folklore speaks of gods coming from the water. Other folklores speak of holy mountains where their old gods lived. These dwelling places may, of course, be sites which were temporarily occupied by "Others" on their visits to earth. Many cultures describe visits from their gods, teaching them agriculture, mathematics, medicine and how to build. These gods could be ETs, ITs, a more advanced human civilisation, or entities from a parallel universe.

It would appear that there may have been human migration as part of the mitigation measures taken by the ETs or ITs to protect humankind from the great flood, resulting from the melting of ice at the end of the Ice Age.

Many cultures start by the sea. Their first towns would have taken advantage of fishing for food, rivers to provide fresh water, and higher coastal rainfall for irrigation. These towns would be under threat from a flood. Underwater towns and cities are now being discovered, on the continental shelves but below the present sea level – Dwarka[174] in India, Pavlopetri[175] in Greece, Atlit-Yam[176] in Israel, and Mahabalipuram[177] in India are such examples.

The underground city at Cappadocia[178] in Turkey can accommodate 20,000 people. It is difficult to imagine this being excavated by humans back in 12000 BCE, but might have been constructed by ITs or ETs to shelter refugees from the floods. Some Native Americans are now desert dwelling, and their folklore describes their being taken underground for shelter by their gods. Perhaps these survivors were originally coastal dwellers and were moved to safety well inland or to higher ground.

iii.   Extra Terrestrials (ETs)

There is a fair amount of evidence to suggest that ETs have visited us in the past. There are drawings and carvings showing entities clad in what appear to be space helmets. However, whilst ITs would not normally need space helmets, it is possible that PUEs might. In fact, almost everything attributed to ETs could, with few exceptions, be attributed to advanced humans, ITs and/or PUEs.

At a primary level, the unique capability required of ETs is that they can travel vast distances in space by using a Wormhole, Gateway or some other technology. It is rightly argued that their civilisations could be billions of years more advanced than humans, and their technology might make this possible. However, the converse need not be true – the presence of visiting entities does not prove that wormholes or some such technology must exist.

It has been postulated that ET entities are altruistically motivated to develop the human species, teaching us many basics, such as farming, mathematics, medicine and astronomy. It is suggested that many of the large-scale building projects of the past were undertaken by, or with their assistance.

It is also hypothesised that their presence could have been motivated by the search for raw materials, particularly gold. The claimed genetic discontinuity[179] in the development of humans could have been introduced by them to generate a workforce. It is interesting that practically every part of human society, worldwide, has always valued gold extremely highly. We also value it today because of the many ways we can use it in electronics. We have been steadily gathering all the gold we can, and storing it for the future. However, there have been persistent rumours that there is gold missing from

Fort Knox, and an audit is proposed[180]. If gold is missing, will the USA ever admit it? Perhaps we have finally made our first payment to the ETs.

h) <u>THEY ARE PRODUCTS OF HUMAN REVERSE ENGINEERING OF "OTHER" TECHNOLOGY</u>

During World War II, Germany was reportedly attempting to reverse-engineer a spaceship which crash-landed there in 1936[181]. There is no record of what happened to it after the war, but it is fairly certain that either the USSR or the USA spirited it away. Given the speed with which the USA started to advance technically after the war, they would seem the better bet.

Two subsequent crashes at Roswell[182] (1947) and Kingman[183] (1953) may have further helped, but the major benefit of these was that there was an "Other" survivor from these crashes, who assisted with the reverse-engineering project for many years.

Although there have been a fair number of whistle-blowers, including Bob Lazar[184], there has never been a public admission that they are reverse-engineering flying saucers. There have, however, been claims that a number of technical advances have been facilitated by this process, including fibre-optics, stealth technology, night vision and micro-processors[185].

It is possible that some of the UAP crashes since the Springfield crash may have been of US prototypes. Scrapped USA jet-powered circular aircraft prototypes[186] were reported in a scrap-yard at MacDill Air Force base in 1967. It is also rumoured that the USA has a way of shooting down UAPs. One possibility is the use of Electro-Magnetic Pulses (EMPs), and some sort of beam weapon[187] has been witnessed firing from Area 52 which is another name for Dugway Proving Ground, a military installation in Utah.

Since then, it would appear that the US has been developing a flying triangle, called the TR3B[188], which does not appear to be jet-powered, but may have been part-dirigible at first. This suggests that the reverse-engineering programme, and the assistance of the "Other" J Rod, has paid dividends. Some flight testing was done in the Hudson Valley in the 1980s, followed by deployments over Belgium in the 1990s as well as over England in 1993. There were many reports of these vehicles apparently following patrol routes down to the east of the Rockies and once over Phoenix, Arizona in 1997, viewed by thousands and known as the Phoenix Lights.

Given that at least 30 years have passed since flying triangles appear to have become operational, there could be, by now, a third generation of US UAPs that we haven't seen yet.

The extremely aggressive way in which some element within the US has attacked eyewitnesses, and UAP researchers, sending its thugs to threaten them and their families, suggests a grim determination to conceal their activities, offering virtual proof in the process. Of course, this may be in response to the agreement between Eisenhower and the Grays, dating back to 1954.

This at least accounts for one group of UAVs.

Chapter 4

WHERE ARE THEY?

a. <u>WHERE ARE THEIR BASES?</u>

i. Elsewhere

To state the obvious, it is inevitable that our Moon should be considered as a primary base, with claims that it is artificial and hollow. There are also claims that there are "Other" facilities on the back of the Moon. The space-going earthly nations will know the truth of this, but they still cling to their charade of denying everything, far beyond what might be considered reasonable.

It is also reasonable to assume that, if there are "Others" on Earth, they may well also have local bases elsewhere in our Solar System.

ii. Underground

One of the most popular proposed locations for an "Other" base is Antarctica. It is suggested that Nazi Germany and the USA found it in the 1930s-40s. It is claimed that Admiral Byrd, during Operation High Jump[189], was taken into an under-ice city to meet one of the "Other" leaders, who lectured him on the perils of the Atom Bomb. Byrd then aborted his mission, dashed back to Washington to report and was ordered to keep silent. It is claimed that his personal diary has been found. There have since been many reports by scientific and military personnel stationed in Antarctica of strange events, sightings and encounters. It should be noted that Admiral Byrd did not comment on the appearance of the "Other" he met. This could be presumed to mean that the person he met was human in appearance, or he would have said something. Antarctica is possibly a base for the earlier earth-human civilisation. They are reported to have flying saucers, and possibly cigar-shaped UAPs like the one which is supposed to have crashed on a glacier on the island of South Georgia. This is a society to blame for some of our UAPs.

When "Others" first made contact with our ancestors, they were described as living in mountains such as Mount Kailash in Tibet, Mount Musiné[190] in Italy, Mount Shasta[191] in California and Mount Denali[192] (formerly called Mount McKinley before reverting to its Athabascan name) in Alaska. There are reportedly many others but some, such as Mount Olympus, do not appear to be occupied any longer. Also UAPs are reported flying into volcanoes such as Popocatépetl in Mexico and Mount Shishaldin[193] in the Aleutian Islands.

Mount Kailash in Tibet is generally associated with a gigantic underground city, and there are present day reports of UAPs flying into it through doorways which appear and disappear. If it is occupied by "Others", there seems to be no evidence to identify what race they belong to, although it is possible that the blue-skinned "Others" from India are there now. However, the sightings are of flying saucer type UAPs, not of the Vimanas, their more conical-shaped UAPs. To live in such a city, they would need to have curbed their argumentative natures. It is interesting that the first representations of Buddha show him as blue-skinned like them, so perhaps they have.

Mount Musiné still gets visits from UAPs, although claimed sightings appear less in number. Mount Shasta in the USA is still a hot-spot of the unexplained. There are reported disappearances, sightings of UAPs, and reports of strangers in the caverns vanishing into the rock walls. These have been described as typical humans, so this may also be a base for the earlier earth-human civilisation. Mount Denali in Alaska appears to be under investigation by the USA. It is claimed that the mountain contains a massive black triangular pyramid, which is generating so much power that it could supply the whole of Canada. It is not known which "Other" race produced this, but it is reminiscent of claims made about the great pyramid in Egypt.

The Falkirk Triangle[194] in Scotland has so many claimed sightings every year, that it is a distinct possibility that there is an UAP underground base in the vicinity.

There is some indication that there is an underground base somewhere in Argentina, probably on the western slopes of the Andes.

Multiple sightings of the smaller illuminated globe type of UAP are reported world-wide, both accompanying larger UAPs and acting in small numbers. It has been hypothesised that these are generally used for reconnaissance although, ever since the World War II Foo Fighters, they are known to have an offensive capability - disrupting electronics just by their proximity. Many of these have been sighted in the vicinity of Skinwalker Ranch in Utah, which has been the subject of intensive investigation for years, and electronic failure is the norm. Native Indians suggest that the Ranch is the site of a portal, and that cryptids have been seen there. There is a spiral of rocks there which they say was constructed as an indicator of a portal. A spiral carved into a rock face is common worldwide and may be an indicator that portals are a global phenomenon. Native Indians suggest that these portals are to Parallel Universes, so it could be that PUEs are watching us closely, without participating actively.

In Dulce[195] in the USA, there is reputed to be a massive underground base beneath Archuleta Mesa, housing up to 18,000 "Others", primarily Grays, in the lowest 3 of 8 levels. In the Dulce area there are reports of UAP sightings as well as cattle mutilations. There was reported in 1979 that there was a fire-fight in the base, probably started by a trigger-happy human, but no-one is sure of the outcome. This could be the base originally agreed in the reported deal with President Eisenhower.

There is deep suspicion that the joint CIA/Australia listening station in Pine Gap[196], south of Alice Springs, has expanded significantly and it is claimed that it may now be a joint USA/"Other" base.

iii. Underwater

It is claimed that about half of UAP sightings are associated with water, in one way or another. Most of these are over land or coastline, and it is suggested that they use water in their propulsion system.

It is perhaps inevitable that there are far fewer UAPs sighted out at sea than there are on land. Of these, most are reported near to the coast by ordinary people, whilst those in the deep sea are mostly by mariners. Sightings by naval personnel tend to be censored.

In 1982, USSR divers in Lake Baikal[197] observed underwater "Others" and made the mistake of trying to capture one. All the divers were suddenly forced to the surface, with three dying from the bends. These "Others" were described as humanoid in appearance, but adapted for life underwater. It seems unlikely that a fully aquatic "Other" would need to evolve legs. They would have to have evolved from amphibians.

There are claims of underwater bases near Cuba, Puerto Rico, Malibu in the USA, and in Lake Fluxian in China. The Cuban and Chinese bases seem now to be uninhabited. The Malibu base could account for all the UAP sightings near Catalina Island and the US Vandenberg rocket base. There have been no claims about what type of "Other" lives there, but they could be ETs or ITs.

In 1956, an "Other" contactee claimed to have been told of three underwater bases, one on the Uruguayan coast 25 miles from Buenos Aires[198], and one 100 miles south of Rio de Janeiro[199]. It was also claimed that there was one somewhere in the Gulf of Mexico.

In June 1959, it was claimed that a UAP emerged from the sea in Tierra del Fuego, Argentina. This was observed on two separate occasions.

In 1963, it was claimed that an underwater UAP interrupted a US Navy exercise[200] off the coast of Puerto Rico.

There are reports of an underwater base on the deep water off the Bahamas, which has power and communication links to an US base, AUTEC, on the Andros Island[201] coast. This could be another example of "Others" working with the US, like in Dulce in New Mexico, and Pine Gap in Australia.

There are possibly two UAP bases in the Japan area. One could be in the mountains in the very north, and one underwater in Micronesia in the south of the Dragon's Triangle

Given that some "Other" UAPs appear to have significant underwater capabilities including being able to enter and pass through the sea at great speed, there would appear to be no great requirement for them to show themselves unless they wanted to, or they wished to leave the planet.

iv. POSSIBLE BASES ELSEWHERE

In many countries, there are not the UAP reporting networks that exist in the US and Europe but, even so it is clear that UAPs and their bases are a worldwide phenomenon.

b) THE BALD FACTS

In North America, there are "Other" bases at:

- Dulce
- Mount Denali in Alaska
- Mount Shasta in the Rockies
- The Rockies in Utah
- Mount Shishaldin in the Aleutian Islands
- In the sea off Marabou
- In the sea off the Bahamas
- Puerto Rico

In South & Central America, there are "Other" Bases at:

- Popocatepetl in Mexico
- In the sea off Rio de Janeiro
- In the sea south of Buenos Aires
- In the Andes in north west Argentina
- In a lake in west Brazil

In Europe, there are bases at:

- The Falkirk Triangle in Scotland
- Mount Musiné in north Italy
- Bucegi Mountains, Romania[202]

In Asia, there is not the information available to identify many bases, but there are "Other" bases at:

- Mount Kailash in Tibet
- Northern Afghanistan
- Lake Baikal in Russia
- Kola Peninsular, Russia
- The Mountains of north Japan
- Underwater in Micronesia

In the Southern Hemisphere, there are bases at:

- Pine Gap, Australia
- Blue Mountains, Australia
- Antarctica

It is likely that this is an under-estimate of the total number of "Other" bases world-wide but, even so, this does seem excessive if all they are doing is stopping us from destroying our planet with atom bombs, gathering rare earths, and studying our genetics.

What we are seeing is colonisation, with multiple "Other" species sharing the planet with us, with only their over-flights intruding on our consciousness.

c) COLONISATION IN SPACE

Some of our wealthier inhabitants are already thinking of colonising Mars, suggesting that this is the only way to guarantee the long-term survival of the human species. I hope that they are really thinking further than this because, if some malevolent "Others" decide that they want Earth for real-estate or humans for slave labourers or food, they could probably deal with both Earth and Mars with one casual swipe.

The answer is to have as many fully autonomous colonies as possible, and to keep them small. This is what is happening on Earth. There are reputed to be over 150 generally friendly "Other" species, and the Russians are supposed to have produced a directory of 50 of these from their observations[203]. If this is so, this provides a rough estimate of the number of small colonies that could be hiding on Earth.

Chapter 5

WHY IS THERE NO NATIONAL OR INTERNATIONAL RESPONSE?

a) THE US AND ETs

The Roswell Crash occurred in 1947. That same year President Truman signed the National Security Act which set up the Department of Defense, the USAF and the CIA. It is claimed that an organisation called Majestic-12[204] was set up as the same time, reporting directly to the President, to deal with all matters relating to UAPs.

In 1949, Majestic's secretary, James Forrestal, died after falling from a 16th-floor window at Bethesda Naval Hospital[205]. While officially ruled a suicide, his death was shrouded in controversy and sparked numerous conspiracy theories about a Majestic power-grab.

In 1952, it is said that Washington was buzzed two weeks in a row by a cluster of UAPs. USA fighters, which were scrambled to intercept, were unable to stop them[206]. This could be taken as a threat to the USA – "Surrender or be Hammered". The story goes that President Eisenhower was then approached by two different groups of "Others" – Nordics and Grays. He eventually made a deal with Grays. This allowed them to establish a base, and to capture earthly cattle and humans for medical experiments, provided that they returned the humans safely home. In return, they agreed to give the US some of their technology. What he didn't realise was that they had another item on their agenda too. It soon turned out that the US was not to be permitted to use nuclear weapons.

It may be that this is the moment when the US surrendered its sovereignty to the ruling Grays. It is alleged that Majestic 12 works vigorously to prevent any word about UAPs from getting out, using thugs to intimidate possible eye-witnesses and their families, adopting tactics which sometimes resulted in fatalities. They instigated a policy of explaining away every reported sighting with sometimes fanciful theories, confiscating any evidence and even denying that whistle-blowers had worked where they claimed. It has now got to the state where they are trying to defend the indefensible.

It appears that the "Others" are setting the agenda, and will not allow any information to be made public until the general population is considered ready for it. The "Others" will decide when that is. Some countries around the world have decided that they are not constrained by the US agreement with the Grays and have released everything about UAPs that they can. In the US, the pecking order is "Others", Majestic, then the President, in that order.

b) POSSIBLE "OTHER" INTERESTS

Some ITs, together with the earlier civilisation if it exists, are probably aware of the position regarding the Grays. The question is whether this suits them. It is likely that any earlier civilisation would already have come into contact with the Grays, and would have a better understanding of its motivations. They may be quite content to let things run their course. The situation for ITs would depend greatly on their location. Any sea-dwelling ITs would probably have come into contact with any sea-dwelling colonists over the years, and would share their concerns. For any deep-underground dwelling ITs, the position would be specific to each case.

c) <u>MAJESTIC'S ACTIVITIES</u>

It has been suggested that Majestic no longer reports everything to the President, depending on whether they consider the incumbent to be to their liking (i.e. Republican or sometimes Democrat). Effectively, there has been a power-struggle which the President had lost.

It has even been suggested that, when President Kennedy proposed sharing space technology with the USSR, they were the ones who arranged his assassination. Lee Harvey Oswald would never have had the chance to kill Kennedy in Dallas, had an assassination plot in Chicago[207] succeeded three weeks earlier, a plot that has been seldom mentioned over the years. Kennedy was due to arrive in Chicago the morning of Nov 2nd to attend the Army-Air Force football game at Soldier Field and ride in a parade. Newspapers had even printed JFK's detailed travel plan from O'Hare airport to the Loop.

Although police were preparing to line the motorcade route, Secret Service officials in Chicago were deeply troubled about the visit because of two secret threats. Right-wing radical and Kennedy denouncer Thomas Vallee had arranged to be off work for JFK's visit; Vallee, an expert marksman, was arrested with an M1 rifle, a handgun and 3,000 rounds of ammo. But then there was a phone call to federal agents from a motel manager that she had seen two Cubans with several automatic rifles with telescopic sights, with an outline of the route that President Kennedy was supposed to take in Chicago that would bring him past that building. Somehow, agents bungled surveillance of those two suspected Cuban hit men. They disappeared and were never identified. No one was even sent to the room to fingerprint it or get an ID. This shows a scenario where one hit man is set up, with two others to do the job - very similar to the Dallas scenario.

Certainly US Congress has now come to the conclusion that there had been a conspiracy to kill him and that there were probably two shooters at Dallas.

It is alleged that when President Carter took office, he was given an official briefing on UAPs and "Others" and that, afterwards he was found in his office, sobbing his eyes out[208]. Obviously, this cannot be verified.

The reverse engineering of UAPs has continued, probably now under the control of Majestic, and the US has now become a two part state. NASA undertakes the public part of the space programme, whilst other agencies finance the private sector to undertake Majestic's space programme. NASA has to be very careful to avoid accidentally releasing anything about Majestic's activities. In these areas, the President has to do what he or she is told.

It has been reported that Majestic's operations are now so advanced that they could leave Earth behind if they wished[209]. One wonders why some billionaires are so keen to develop their own spaceships, and go to Mars.

d) RECENT EVENTS IN THE USA SPACE PROGRAMME

It is to be hoped that Majestic is playing a very subtle game and not desperately clinging to power at the expense of humanity. The "Others" may be in control of the US, but it may be that they are not aware of the activities of Intra Terrestrials, the more advanced human civilisation, or PUEs. Many of Majestic's actions would need to be kept out of sight of "Others". The appearance of the TR3B over centres of population worldwide, showing as many lights as possible, could be a form of miss-direction both for humanity and ETs, diverting attention from any new vehicles being developed. The US has done this type of thing before.

The English hacker, Gary McKinnon[210] whose extensive search through US computer networks was allegedly conducted between February 2001 and March 2002, gained access to Excel spreadsheets on NASA computers. He says, one was titled "Non-Terrestrial Officers." It contained names and ranks of US Air Force personnel who were not registered anywhere else. This is well before the creation of the US Space Force in 2019.

The spreadsheet also contained information about ship-to-ship transfers, but he'd never seen the names of these ships noted anywhere else. He says the ships were titled **USSS (Unites States Space Ship)**. The US authorities tried to charge him and extradite him, but the UK has finally rejected this. The charges and proposed penalties were so severe that it would appear that the US authorities were really out to get him. Perhaps he had found something important.

First launched in 2010, the Boeing-made, automated reusable space plane X-37B, has spent as long as 908 days in space at a time. Its 29 feet (9 meters) long with a wingspan of almost 15 feet (4.5 meters).

Early in the Space Shuttle programme, the US admitted to having a cadre of 20 Manned Spaceship Engineers[211] (MSEs) who worked with NASA on classified projects such as launching spy satellites. After the 1986 Challenger accident, as NASA struggled to return the shuttle to flight, the Air Force sped up its plans to move payloads back to unmanned rockets. By 1988—the year NASA returned the shuttle to service—it was claimed that the MSE cadre had been disbanded, its members scattered to new assignments. (Of the 27 officers in the first two MSE groups, five would later become generals.) The US Space Force was established December 20th 2019 when the National Defense Authorization Act[212] was signed into law, creating the first new branch of the US armed services in 73 years.

Given the way governments work, the announcement that something has been cancelled doesn't necessarily mean that it really has been, so there was probably some form of secret organization in operation for the whole of the 1988-2019 period, as suggested by the NASA spreadsheet described by Gary McKinnon. When the US military's classified mini space shuttle X-37B[213] returned to Earth in March 2025 after circling the world for 434 days, this was the 7th mission of the type. The space plane had blasted into orbit from NASA's Kennedy Space Centre in December 2023 on a secret mission. Launched by SpaceX, it was claimed to carry no people, just military experiments.

This vehicle did not suddenly magically appear in 2019 when the US Space Service was created. It first flew in 2010. Many forget that earlier versions of this Boeing craft also had a secret military role. Although ostensibly a civilian program, they conducted a series of missions from 1982-1992 on behalf of the National Reconnaissance Office, launching a series of classified spy satellites, and were probably not sitting idle in the 1992 -2010 period.

e) THE PRESENT SITUATION

It would appear that no-one is doing anything about the presence of UAPs and "Others", because there is nothing that they can do. Some of these colonies have been here since the dawn of our civilisation, and may even be the cause of it.

The Anunnaki appear to have initiated the drive to bring humans up to a level where they could be accepted by "Others". They have now taken a back seat. Until the start of the industrial revolution, Earth was a delightful rural idyll for any colonists, but we started some serious pollution about a hundred years ago, and then we started playing with atomic weapons.

Needless to say, the various colonists on our planet were not happy with this on two accounts: they didn't want us to destroy the planet, and they didn't want us signalling to malevolent "Others" that we were there. The Grays then stepped in and, along with some "Others" of a similar mindset, such as the Nordics, are starting to take control. They have certainly taken steps to prevent the use of nuclear weapons, and are keeping a close eye on our other nuclear facilities, having even stepped in to mitigate the effects at Chernobyl[214]. It remains to be seen whether they intend to do anything about the pollution we are causing.

They are probably aware of the development of the US's space capability, and may even be moving towards co-ordination of activities. The Pentyrch Incident on 22nd February 2016[215], in the UK, shows that some cooperation occurs. Who else could have tipped off the UK government about the expected arrival of a hostile UAP? The US has responded by keeping information away from its own population and, in general, other countries are going along.

Those amongst us who have been campaigning for full disclosure by governments of the existence of "Others" and UAPs, may be unaware of the colonial aspect. Whilst the general population could probably cope with an admission to the former, the knowledge that we are really a colony planet, may be harder to take.

Chapter 6

OUR GUARDIANS

a) GENERAL

The universe is a big and cruel place. We have only come into serious contact with the more benign species of "Others". There are probably also malevolent species, and the "Others" that we know have apparently been shielding us from them.

There are reports of buildings in Siberia which have been nicknamed "Cauldrons", and which seem to have the capability to destroy aerial objects such as in the Tunguska incident in 1908. It is possible that the Grays have installed a worldwide network of these, and they could be powered from the mysterious pyramid inside Mount Hayes in Alaska. The oral history in Siberia speaks of these devices being used on multiple occasions in the past.

b) AERIAL BATTLES

In the past, there have been battles in the skies, which have been reported by human spectators. There is written record of a day-long battle over Nuremberg[216], in which many UAPs were destroyed, and another over the Baltic in 1665. Our defenders have been putting their lives on the line to preserve our planet.

It must have been a bitter pill to swallow, when we began trying to shoot them down ourselves. Nevertheless, they seem tolerant of our ignorance and, still appear determined to protect us.

As already described, a fleet of some 200 UAPs was seen in broad daylight launching from their Catalina base in 1992[217]. There may have been more launched from other bases at the same time, perhaps unseen because of it being night-time there. Given that the "Others" are generally wary of giving away their presence, and prefer to operate at night, this must have been a serious emergency. It rather sounds like a scramble for battle.

The Pentyrch incident[218] in the UK in 2016 is evidence of the Gray's cooperation at government level. Reports suggest that the RAF had received prior information of an unwelcome UAP, and were waiting for it. As soon as its arrival was detected, it was attacked, and two large explosions were heard.

c) <u>IN SUMMARY</u>

Yes, there are UAPs in our skies, and yes "Others" exist. Most of these UAPs are either small un-crewed reconnaissance drones or are crewed by Grays. They appear to be fighting a desperate battle to save our planet, even at risk of their own lives.

Of course, it is not just humans that they are trying to save. There are also an unknown number of "Other" colonies in our mountains and under the seas.

PART 2

WE HAVE 29% - ET HAS THE REST

THE UFO, E.T, ALIEN TRILOGY

THE UFO, E.T, ALIEN TRILOGY

Table of Contents

	PREFACE		85
1	INTRODUCTION		87
	a	EARLY TIMES	87
	b	THE GODS ARRIVE	88
	c	OR WERE THEY HERE ALREADY?	91
2	WHERE DID INTER-TERRESTRIALS COME FROM?		93
	a	GENERAL	93
	b	FISH	93
	c	CETACEANS	94
	d	CEPHALOPODS	95
	e	ET ALIA	96
3	WHERE ARE THE MOST LIKELY BASES?		98
	a	GENERAL	98
	b	INDIAN OCEAN	102
	c	PACIFIC OCEAN	105
	d	ATLANTIC, CARRIBEAN & THE BALTIC	121
	e	INLAND WATERS	136
4	WHAT "OTHERS" DO WE KNOW ABOUT?		140
	a	GENERAL	140
	b	WATER MONSTERS	141
	c	GIANT APES	143
	d	THE GRAYS	144
	e	TALL WHITES	145
	f	NORDICS	146
	g	DWARFS & THE LITTLE PEOPLE	147
	h	WINGED CREATURES	149
	i	LIZARD PEOPLE	150
	j	CATS & DOGS	150
	k	CHUPACABRA	151
	l	TELOSIANS	152
	m	A WARNING TO RESEARCHERS	152

5	WHERE ARE THEY HIDING?		153
	a	BASES V. COLONIES	153
	b	COLONIAL ATTITUDES	155
	c	COLONIES	156
6	CONCLUSIONS		159
	APPENDIX		161

LIST OF FIGURES & TABLES

	CAPTION	PAGE
Figure 1	Map of the world showing sea levels and ice sheets at the peak of the last ice age	89
Figure 2	The seabed of the Pacific Ocean	106
Figure 3	Topography of the seafloor of the Caribbean Sea	125
Table 1	A summary of the land-based "Others" sightings reported in this book	101

PREFACE

In the first part of this book "UFO – Friend or Foe?[219]", I made a broad-ranging sweep through the history of sightings across as much of our planet as I could.

In reviewing evidence of all the sightings of UAPs, I was easily able to convince myself, and hopefully my readers, that they really existed, and that they had some bases underground, and underwater. There could easily be 20 or 30 such bases, judging from the distribution of UAP sightings. It is likely that this is an under-estimate but, even so, this does seem excessive if all they are doing is stopping us from destroying our planet with atom bombs, gathering rare earths, and studying our genetics.

What we are seeing is colonisation, with "Other" species sharing the planet with us, with only their over-flights intruding on our consciousness. The answer to malevolent "Others" is to have as many fully autonomous colonies as possible, and to keep them small. This is what is happening on Earth, although it is not clear how many are involved.

So there is every chance that many of the underwater bases could belong to Extra-Terrestrials (ETs). They may be water-breathers, but there is no reason why they should not be species requiring a gaseous atmosphere, living in an artificial biosphere. Thus they could also be other earth-evolved species – Intra-Terrestrials (ITs), members of an earlier human race, or even entities from another parallel dimension (PDEs).

Chapter 1

INTRODUCTION

a) <u>EARLY TIMES</u>

The last ice-age started about 2.6 million years BCE and finished about 11,700 years BCE. It must be remembered, though, that a large proportion of the land-mass of our planet was not covered by ice-sheets, although it was a lot colder. There are indications of civilisation going back in Gobekli Tepe[220] in Turkey to at least 12,000 BCE, but there are certain features, such as carvings of figures with six fingers, which suggest they may not have belonged to our current human civilisation.

The first signs of our present human civilisation are thought to occur at roughly 6,500 BCE in Sumeria. There are legends of gods called Anunnaki[221] at that time, teaching medicine, agriculture, animal husbandry, astronomy and mathematics. The wheel was first adopted about then.

These gods are the first recorded contact with "Others" by our present civilisation. Of course this does not mean that this was the first arrival of these "Others" on our planet. Indeed, there must have been at least one prior arrival if the hypothetical genetic anomaly[222] in our evolution is to be believed.

However, as I aimed to demonstrate in my previous book, "Others" have been colonists here, perhaps for millennia, living on mountain tops, in caves, or in artificial bases underwater in lakes or the sea.

The reason I have already mentioned the ice-age is that the vast northern ice-sheets could have determined, to some extent, where these colonising "Others" may have set up camp. (See Figure 1)

Even within the ice-sheets, there would still have been massive mountains rising above the ice, offering potential base sites.
In the lands away from the ice sheets, they could have come at any time. There would have been plenty of choice of mountains, lakes or oceans, even thought the sea level would have been perhaps 120 metres lower[223] then.

b) THE GODS ARRIVE

I am going to be concentrating in this book on underwater bases, although there are certainly still some mountain bases such as Mount Hayes[224] in Alaska, Mount Shasta[225] in California and Mount Kailash[226] in Tibet as well as volcanoes such as Mount Shishaldin[227] in the Aleutian Islands and Mount Popacatepétl[228] in Mexico,

If some early colonists assisted in the development of various human tribes, they could have come from the oceans.

There are many oral histories of "Others", taken then to be gods, visiting tribes to teach such basic things as agriculture, cooking, fishing, medicine and building.

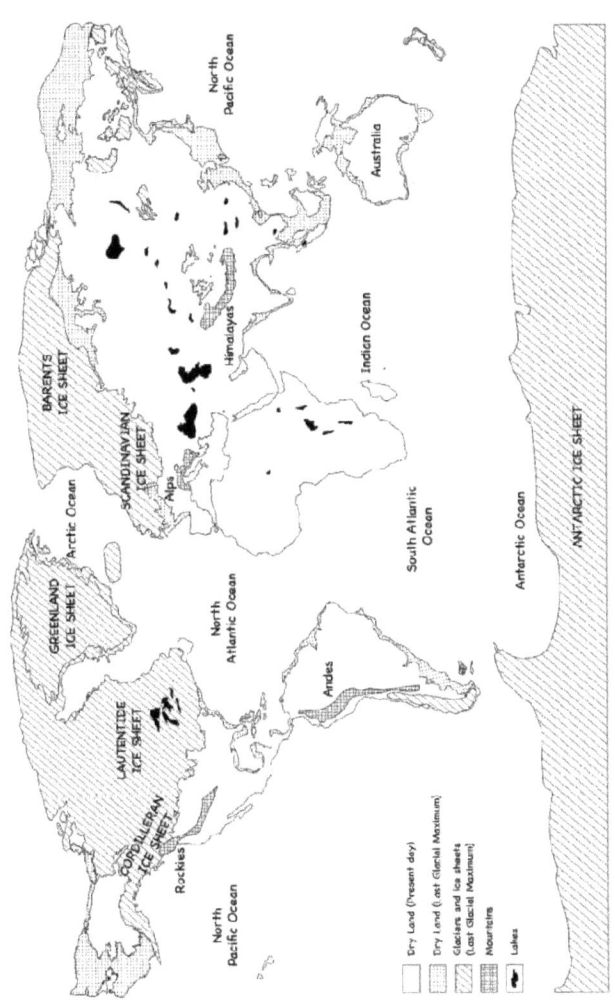

FIGURE 1. Map of the world showing sea levels and ice sheets at the peak of the last ice age.

Some of these are specifically described as sea gods, even though they may have arrived in flaming chariots:

- **Oannes**,[229] in Mesopotamian mythology, was an amphibious being who taught mankind wisdom. He had the form of a fish but with the head of a man under his fish's head and under his fish's tail the feet of a man. In the daytime he came up to the seashore of the Persian Gulf and instructed mankind in writing, the arts, and the sciences.

- **The Nommo**[230] are ancestral spirits (sometimes referred to as deities) worshipped by the Dogon tribe of Mali, Africa. The word Nommo is derived from a Dogon word meaning: "to make one drink". Folk art depictions of the Nommos show creatures with humanoid upper torsos, legs/feet, and a fish-like lower torso and tail. The Nommos are also referred to as "The Masters of the Water", "The Monitors", and "The Teachers". The Dogon legends state that the Nommos required a watery environment in which to live.

- **Dagon**[231] was the god of the Philistines. The idol was represented in the combination of both man and fish. The name 'Dagon' is derived from 'dag' which means 'fish'. His name is a lot like 'Dogon'. Dagon was worshipped by the early Amorites, by the people of Ebla, by the people of Ugarit and a chief god (perhaps the chief god) of the Biblical Philistines.

- **Greek Sea Gods** It would appear that, as well as Titans, the Greek gods included many Sea Gods such as Amphitrite, Poseidon, and Triton as well as countless Sirens and Nymphs.

I think it is significant that the claimed Russian KGB list of "Others", which has supposedly been translated into English, makes no mention of amphibious aliens, even though there are records of possible sightings in Russia in other reports.

In 1982, in Lake Baikal[232], Russian divers came across underwater "Others" and attempted to capture them. This resulted in the deaths of three of the five divers. The "Others" were described as basically humanoid but adapted for the water.

In another book[233] listing "Other" races, there are references to several kinds of amphibians. There is one which is very similar to humans, but with webbed hands and feet. A second is more reptilian, and is believed to have returned to the sea after being land-dwelling. This book also refers to the Alabrans, who need to hydrate themselves regularly.

c) <u>OR WERE THEY THERE ALREADY?</u>

Although the great flood, which is recorded in the Bible and other east Mediterranean texts, is generally assumed to be the Mediterranean Sea breaking into the Black Sea, there is folklore of a Great Flood, sufficient to destroy all human society, in many human cultures around the world, and they would never have even heard of the Black Sea.

There are submerged coastal cities dated from the end of the ice-age, which are considered to be proof of this. As all the ice melted, it raised the sea level, leaving these cities under-water. However, the amount of water, necessary to destroy all human society would have far exceeded the melt-water from the end of the ice-age, so some have suggested that perhaps it came from some extra-terrestrial source such as water-filled planetoids. In this case, where did it all go to afterwards?

These submerged cities have been attributed to a part of the human race which developed earlier than us, and which is now called the Telosians. Alternatively, construction of these cities is laid at the feet of "Others", both ETs and ITs, and to the present human race. It reality, it is probably a mixture of all four.

As these cities, which were originally on dry land, were being destroyed, it resulted in considerable disruption to the stumbling development of humanity. This was probably the opportunity for "Others" to move in unobserved.

The end result is that, underneath our seas today, there are a number of colonies, hidden from human sight, where various "Others" can now live out their lives. These are:-

- Benevolent "Other" military bases, focused on protecting Earth from malevolent "Others" and, later on, preventing humanity from miss-using its atomic capabilities.
- Small "Other" colonies, focused on forming micro-versions of their home planetary societies to preserve their species in case of catastrophe.
- Intra-Terrestrials, living in isolation from their land-based neighbours.
- Perhaps residual populations of that earlier human race, the Telosians, who are claimed to be the descendents of Lemuria, a sunken land similar to Atlantis. However, some claim that they were wiped out, millennia ago.

Chapter 2

WHERE DID THE INTRA-TERRESTRIALS COME FROM?

a) GENERAL

The aquatic dinosaurs supposedly did not survive the land dinosaur extinction event. It has been suggested that the seas became too acidic for them. However, it is evident that other aquatic creatures did survive – the sea dinosaurs had to eat something, and not all these "somethings" can have disappeared.

b) FISH

To survive predation by the aquatic dinosaurs, one can speculate that it would have been an advantage for fish to have the ability to plan and to communicate with others of the same species. Some species could have developed more intelligence that we give them credit for. Survival has always been a driver for development.

The top of this food chain are the sharks. They are fish whose skeletons are made from cartilage rather than bone. They are part of a group including skates, rays, sawfish and over 350 species of shark.

We know of species of these which demonstrate remarkable co-operative hunting abilities and the most intelligent might have learnt how to avoid humans entirely as a survival technique. We may never know when or if they developed abilities beyond our own.

c) CETACEANS

In his book "The Naked Ape[234]", Desmond Morris suggested that, at one stage, early in their development, humans dwelt on shorelines, foraging for fish, shellfish, seaweed etc. To offer proof, he points out that what hair we and some water mammals have on our bodies is very different from other animals. It is streamlined from head to toe, perhaps to offer less resistance when swimming.

If humans were one-time beachcombers, it is not beyond the realms of possibility that two strains diverged here, leaving no fossil record. Some strains opted to remain in the water, remaining oxygen-breathing mammals, and evolving into dolphins, porpoises and whales.

We know that some species of cetaceans are very intelligent, and one can speculate that one deep-diving group might have become far brighter than us. Some whales can dive extremely deep, and stay under for long periods. They may have managed to avoid us entirely.

In the 1950s, John Lilly's[235] research helped to advance the view of dolphins as intelligent and socially aware creatures, a perspective that was a marked contrast to earlier, negative views of them. They have been shown to be quick learners and innovative, capable of teaching skills to others.

Even if they do not possess hands to let them become technologically developed, who can say how far their mental abilities could have advanced?

d) CEPHALOPODS

A cephalopod is any member of a class including squid, octopus, cuttlefish, or nautilus. They are widely regarded as the most intelligent of the invertebrates and have well-developed senses and large brains. The nervous system of cephalopods is the most complex of the invertebrates.

The sequencing of a full cephalopod genome has remained challenging to researchers due to the length and repetition of their DNA[236]. The genome showed similar patterns to other marine invertebrates with significant additions assumed to be unique to cephalopods. Cephalopod genomes are generally large, with some, like the squid, being larger than the human genome. Studies have revealed large-scale genomic rearrangements, which likely contribute to cephalopod morphological innovations.

Research has shown that octopuses can form concepts, plan for the future, generate a cognitive map, self-monitor for apparent pain and manipulate communication.[237]

Scientists have had great difficulty in fitting these creatures into any family tree, and it has been suggested that this is because they did not originate on Earth, although they may have been here for millions of years.

Nevertheless, they are here now, are extremely intelligent and have the ability to manipulate their environment. Not needing to breathe air, they could dwell deep underwater and could be an IT species, which we have never detected.

e) <u>ET ALIA</u>

If humans managed to evolve from small underground creatures in the time since the extinction of the dinosaurs, then it is not beyond the realms of possibility that other mammals, or even birds, which are descendants of dinosaurs, did as well, particularly if they had a running start on us.

Equally, there are creatures like reptiles, such as crocodiles, alligators, lizards and iguana still alive which, although not descended from dinosaurs, are historically related to them. It might be that some of their relatives could have out-evolved us.

The only thing missing is an appropriate fossil record. Perhaps we will find one one-day. To some extent, this will depend on whether individual species were capable of developing manipulatory appendages. If not, perhaps the species is in plain sight today, with unrecognised mental capacity. For now, we are specifically interested in sea dwelling creatures, but who knows?

In reality, before they can occupy our oceans undetected, ITs need either to be able to breathe underwater, or to have the ability to construct underwater habitats. That is not to say that ITs cannot occupy hidden places on the land. Native Americans talk of underground Ant People[238], and these are represented in petroglyphs in many places around the world.

Perhaps the best bet for an underwater IT is a cephalopod. The ones we know are very intelligent and equipped with more manipulatory appendages that we have. Who can say whether there is a more technically advanced version of these living in the deeps?

Beyond that, any amphibian or non-hoofed mammal could evolve into an advanced technological species, capable of building its own underwater habitat.

Chapter 3

WHERE ARE THE MOST LIKELY BASES?

a) <u>GENERAL</u>

One place to start is to look at the frequency of submersible UAP sightings and coastal sightings of "Others". In the first volume of his magnum opus, Richard Dolan[239] looks at all the credible sightings of USOs up to 1969 – a total of 178. His future two volumes will cover the remaining 494 up to the year 2024. For the moment, I will have to confine my analysis to the data in volume one and, where possible, combine this with data from other sources.

I have chosen to exclude certain classes of sightings from my analysis. First and foremost, I have virtually always excluded sightings where there is only one witness. Also, it has been claimed that UAPs hide inside clouds and volcanic smoke, but these dark spots could easily be denser patches and, too often, what are clearly lenticular clouds are claimed to be UAP mother-ships. Where sightings consist of spotting a single stationary night-time light, these are the most frequently dismissed as miss-identifications.

Richard Dolan's analysis of each sighting includes date, location and credibility, as well as what other facts he can glean from the sighting report. His credibility rating is particularly valuable, and is sadly lacking from many other sources. He breaks the credibility rating down into:

    Weak   Can think of easier explanations

    Moderate

    Solid

    Strong

    Certain   Absolutely no doubt

He does council caution in looking at UAP sightings, however. He describes in another book[240] how in Illinois in 2008, he spent 2 hours looking at the sky using high quality night-vision binoculars. In that time, he reckoned he saw about 100 high-flying UAPs. He doesn't attribute these just to "Others". He suspects that many were advanced US craft, stemming from their reverse-engineering project.

Even so, I will start my analysis by looking at the reported sightings near to recognized water-based hot spots, and seek to identify any others that stand out. This analysis will become more accurate when Richard Dolan publishes his later volumes.

It must be born in mind that the last ice-age ended about 10,000 years ago, and the folklore of many races speaks of being visited by "Others" before then. At that time much of North America, Northern Europe and Russia was under vast ice-sheets, with only mountains protruding (See Figure 1). Whilst it is not impossible that "Others" established colonies in the lands under these ice-sheets, it would have been easier for them to start with those areas which were not frozen, and with mountain tops.

At the same time, I can look for any evidence of the type of "Other" species occupying the colonies.

I have broken down my analysis by maritime zones. The data is collated in tables in the Appendix.

In the Table on the next page, I have summarised the types of "Others" encountered.

DESCRIPTION	NAME	ORIGIN/ NAME	LOCATIONS	ACTIVITIES
2.0m Tall & Hairy	Giant Apes	Yeti	Worldwide	Solitary existence in forests
1.3m-1.5m Big Head	Small Gray	Reticulum	Sweden & USA	Abductions & Medical Tests
1.6m Gray	Tall Gray	Reticulum	Sweden & USA	Assists Small Grays
2.5m Thin White	Very Tall White	Arcturian	South America & Worldwide	General Peacemaker
2.2m Fair Hair	Nordic	Pleiadians	South America & Worldwide	Helping humans
2.2m Fair Hair Webbed hands	Nordic	Cassiopeians	Northern Asia	Abductions
1.2m - 1.3m	Dwarfs		Wide spread	Many sorts Very ancient
0.1m – 1.0m	Little People		Wide spread	Many sorts Very ancient
2m Black Bird	Mothman		Widespread	Solitary
2m Bird	Avian		North Japan	Ancient
Lizards			Underground	Abductions
Tall Red/Orange			South America	Unknown
No necks			South America	Unknown
Cyclops			Americas	
Cats	Felines	Urmahs	Australia	
Dogs	Canines		Australia	Thylacine
Chupacabra	Goatsucker		South America	"Others" or "Others pets?"
Human 1.8m	Earlier Humans	Telosians	Mt Shastna, Antartica	Unknown

Table 1 A summary of the land-based "Others" sightings reported in this book.

b) INDIAN OCEAN

i. Persian Gulf

This is the nearest potential undersea site to Sumaria the cradle of early human civilisation, where the Anannaki are reported to have nurtured our early civilisation, along with the Avians[241]. It is claimed that the Anunnaki were responsible for the genetic anomaly[242] in human development, 40,000 years ago, by introducing what is called the D allele to human DNA[243].

It is interesting that all the sightings up to 1956 are of this so-called pinwheel. This has not been fully explained but is reported as an under-water light effect, where beams of light seem to radiate from a central point and all stop at much the same distance from this. It is thought that it may be a bio-luminescent effect caused by sound waves, though no-one knows the source of the sound waves. A similar effect is reported in the Malacca Straight in Malaya and the Gulf of Carpentaria in Australia. These are all tropical seas.

I have been unable to find records of any sightings of any sort in the Persian Gulf subsequent to this, so I await Richard Dolan's next volume with interest.

It is claimed that the Anannaki left Earth about 5,000 years ago, presumably leaving their base. This would have either been in the Persian Gulf or in Iraq or Afghanistan.

In my first Chapter, I refer to The Oannes, an "Other" species that lived in the Persian Gulf and came ashore to teach in what is now modern-day Pakistan. They had the form of a fish but with the head of a man under a fish's head and the feet of a man under a fish's tail.

Perhaps they were the occupiers of any Persian Gulf base, or moved in once it had been vacated.

ii) India

These sightings all come from the one book[244]. There is plenty of variation in "Other" species to assess:

- Giant Bird   1
- Yeti   7
- 1.7m   3
- 1.3m Hairy   4
- 1.0m   4

The giant bird could be an Avian but, as they do not appear to range widely, it is more probably one of the black monster variety. The number of Yeti is high, but they are close to where they are said to have originated. What is surprising is that it is claimed that two were caught in 1975. That is a very rare occurrence. The 1.7m hominid is difficult to identify without more information. The 1.3m hairy beast is probably a type of Dwarf, and the 1.0m creature could well be a type of Little People. There are apparently no Grays in this sample.

There are only 4 types of UAP represented:

- Cigar   2
- Disk   6
- Boat (?)   1
- Sphere   1

One of the disks was described as gigantic.

There is not the evidence to suggest that a base or a colony has being set up on the Indian Ocean coast.

c) PACIFIC OCEAN

i. Eastern Russia & the Sea of Okhorsk

The Sea of Okhorsk, Far Eastern Russia, and the Kamchatka Peninsula form a distinct unit. See Figure 2.

Back in the last ice-age, the combination of lower sea-levels and the expanding North American Ice -Sheet, resulted in a link between east Russia and west Alaska. Early humans were able to cross via this link, to populate North America, and possibly even travel further south.

Sightings in the table above come from 3 sources[245,246,247].

From the sightings reported so far, the Sea of Okhorsk, seems particularly active, and certainly the Russians[248] believe there to be at least one "Other" base or colony in the area, possibly in Lake Kronotsky in Kamchatka. The largest UAP, a 200m cigar, was headed in this direction.

The local sightings of 2m "Others" suggest that the web-fingered Nordics, the Cassiopeians[249], have spread to this area from the lakes where they were first reported.

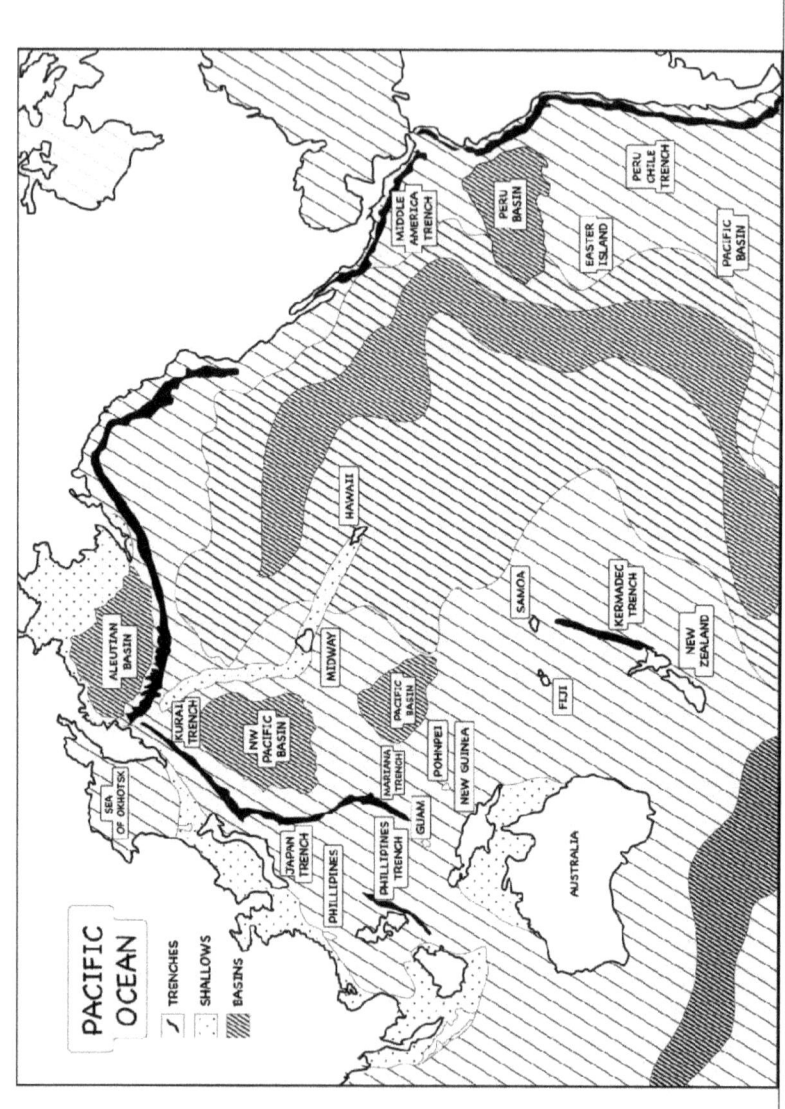

Figure 2 The seabed of the Pacific Ocean

ii. <u>Coastal British Columbia, Washington, Oregon & Southern Alaska</u>

It is necessary to divide Alaska into 2 segments to do it justice. The mainland is huge and seems to contain at least two hot-spots in the mountains, and the south eastern "Inside Passage" archipelago more comfortably belongs with the western lakes and islands of British Columbia, then down into Washington and Oregon.

All the Ketchigan sightings for British Columbia reported above come from a book by Debbie Zieglmeyer[250]. Someone there was keeping records and has passed them on, whilst presumably no-one was recording in the rest of the archipelago during the period from 2000 onward. Other sightings are from George Mitrovic[251].

Debbie also draws attention to the US Navy's SE Alaska Acoustic Measurement Facility (SEAFAC) which is just north of Ketchikan, and points out that there is a possible underwater anomaly in Prince William Sound[252], further north between Juneau and Anchorage. Is this another example of US and "Others" working together like in Dulce, the Bahamas and Pine Gap in Australia, or is SEAFAC just keeping an eye on them?

Considering the number of UAP sightings recorded in these states (35, plus clusters) there have been surprisingly few "Other" sightings. There were 7 disks, the preferred vehicle of the Small Grays, with 3 confirmed sightings of the creatures themselves, together with a Mantis which is always in charge of the Grays when present. The majority UAP, a sphere, was confirmed to have been observed at least 14 times, but this design cannot be associated with any one particular species of "Other", although Small Grays have shown a preference for disks.

Interestingly there were also sightings of a one-eyed giant, or Cyclops, and a Chupacabra, the voracious farm livestock eater.

iii. <u>Coastal California</u>

These records come primarily from three sources[253,254,255]. Certainly, following the nuclear detonations of 1945, we can see an increase in UAP activity in the south of California, and also in the very north, with multiple explosions and UAPs entering the waters around the Catalina Strait. It is likely that a new "Other" base was being constructed somewhere near Catalina Island.

The experience in California seems to fall into three phases. First of all, UAPs arrived around the time of the first atom bomb, and were seen over the whole State, then came the local explosions, combined with further UAP sightings and a few abductions. Finally, in the 1970s, things settled into what was almost routine, with more abductions and fewer reported sightings.

Decade by decade, reported abductions increased:
- 1940s   4
- 1950s   7
- 1960s   13
- 1970s   19
- 1980s   20
- 1990s   16

There are seldom going to be multiple witnesses to abductions, and little subsequent evidence unless there have been devices implanted,

or the abductees are subjected to the questionable process of hypnotic regression. The problems with reported abductions are manifold:
- How many imagined it?
- How many elected to report it?

And most importantly:
- How competent was the hypnotist who helped them remember?

One of the distinctive features of many abductions is that the victim is told by the "Others" exactly what they want to hear:
- "Don't be afraid, I won't hurt you"
- "You've been selected for something special"
- "Something special is going to happen soon"
- "I have a message especially for you"

The victims often seem to have suppressed memories of Grays and Mantis-type "Others". This could be simply the spurious outcome of leading questions by a prejudiced hypnotist but, assuming it is not, the Small Grays seem to be running riot here, clearly indicating the probable presence of at least one Small Gray base nearby.

iv   West Coast of Mexico & Central America

All these sightings come from the one book[256]. Here, there is one single-witness sighing which needs noting. In February 1995, a lorry-driver claimed to have found an "Other", 0.30m tall, sitting in his passenger seat. This individual said that he lived in a colony at the bottom of Lake Chapala, which is near to Guadalajara. It is a very big lake (75km x 25 km), but shallow, with a maximum depth of

9.0m. He claimed that he was descended from the occupants of an UAP that crashed thousands of years ago. This would certainly fit in with the small people described in Mexican folklore, that they called Chaneques.

There were some puzzling sightings in this table with, in one instance, a group of 18 human-height "Others" and, on another occasion, a group of 14 tall "Others". No-one felt inclined to stop them to ask what was going on, but it certainly seems strange even by UAP standards. If these two groups are excluded from any assessment, "Others" shorter than 1.0 metre high are the predominant entities, suggesting that there really is at least one "Little People" colony somewhere in the area. Given that they have also been observed on the other coast of Mexico as well, they seem to have made themselves at home.

v   Peru

All these records come principally from the one book[257], and, I have also included a few aircraft sightings from the book by John Scott Chace.[258]

Sightings here seem to relate to population density, with many occurring in the Lima area or along the coastline. In particular, there is no evidence here of any preponderance between the coast and Lake Titicaca. Single-witness sightings claim there is an underwater base or colony there, but there is insufficient evidence to support this.

As usual, there is a mix of species contributing to the "Others" sighting list:

- Very Tall 2.2m   11
- Very Tall Copper hair   2
- Tall 2m fair hair   5
- 1.6m Egg shaped head   1
- 1.5m human-like   3
- Dwarf   1
- 1.0m Cyclops   10
- 0.8m greenish   8

As in Argentina, the list is dominated by 2.2m "Others", and very small "Others" of one sort or another. The sighting of a Dwarf could be misleading, because it could be one of the smaller ones. The one Gray sighting shows that they are, as ever, sniffing round for any opportunity.

The 0.8m species seems to be a bit frog-like and do not walk in the same manner other humanoids do, These could be the creatures known in local folklore as Karka, possibly an onomatopoeic name suggesting that they might have evolved on Earth and are what are known as Intra-Terrestrials.

The Cyclops-like "Others" have been described as having features which might be adaptions for a maritime existence, such as fin-like hands.

Unfortunately there is not the information to indicate where any "Other" bases or colonies might be. There is insufficient evidence to suggest one in or near Lake Titicaca, but there is an underwater trench running close to sea-shore which could, as in Argentina, provide a suitable location for a base or colony.

vi.   Chile

Most of these records come from the one book[259]. Details of military and aircraft sightings come from another by John Scott Chace[260]. In Chile, there appear to be very few claims of abductions.

There are a few occasions where there has been a single sighting of a particular species of "Other". These are a Shape-Shifter, a Tall Hairy creature like a Bigfoot, what was probably a Chupacabra, a Cat-like creature and, of course the ubiquitous Small Gray. The more frequent sightings have been:

- Tall over 3m   9
- Over 2m no neck   1
- Over 2m Nordic   16
- 1.2 m   6
- Large Bird   4
- Less than 1m   6

The Nordics are often described as Angelic Beings. The Birds appear to frighten witnesses but do not appear hostile. Chilean folklore describes little people living in the forest, that they call Trauco. This may be historical proof of their presence in Chile for long enough to enter into folklore, something which doesn't happen overnight.

The dominating presences are the extremely tall "Others" and the Nordics.

There is no indication from the distribution of sightings along the coast that there is any sort of underwater base or colony. The most sightings are in the wider Santiago area, so perhaps there is an underground base inland there, accommodating the very tall whites and the Little People who are often seen together. There is an underwater trench off the coast of Chile which could also provide a venue for a base.

vii.   New Zealand

All these records of sightings come from one book[261]. There seem to be far more sightings for these islands than for much larger countries. I suspect that this is due to diligence in recording rather than actual frequency.

Strangely there was a sighting of a large predatory cat, although New Zealand has no large cats amongst its fauna. Yeti-like creatures have also been sighted on several occasions. There were also a number of sightings of weird creatures in the sea, getting caught up in fishing nets. I suspect that the sighting of a giant 20m turtle is more likely to be the top of some form of UAP.

The early reports of airships, and then cigar shaped UAPs, all over North and South Island suggest a detailed reconnaissance of the islands during 1909. It would appear that the decision was made that this was a suitable place for a colony and, since then, one thing which stands out is the number sightings in or near Auckland on North Island. As Auckland lies at the south end of an undersea trench, it is certainly credible that an "Other" colony is now sited somewhere there.

Although "tall", "human-like", "dwarf" and "very small" entities have been sighted, there appear to be no records of Small Grays. There is insufficient evidence to deduce the "Other" species, with any certainty, but human-like entities are in a slight majority, so perhaps the earlier human civilisation, the Telosians, has set up home here again.

viii. Australia, Papua & New Guinea

The listing includes coastal sightings. Indeed, other than the Pine Gap facility near Alice Springs, there are very few sightings inland.

Most of these sightings are recorded in the one book,[262] with additional material from two others.[263,264] In this listing I have only included a few of the many sightings. I have listed all the UAP sightings in 1954, to show typical frequency in Australia. Sadly, in general, many of the incidents have only a single witness, so I would normally discount them. It is interested to note that many of the abduction events claimed by single witnesses have involved what are known as Small Grays.

There have been many sightings of strange creatures over the years. There were umpteen sightings of Hairy Monsters (Yowie), Waterpuppies which are perhaps freshwater seal, Bunyip, whatever they are[265], and Thylacines[266] such as the reputedly extinct Tasmanian Tiger. Single witnesses, whose reports I do not usually include, have claimed sightings of "Others" as small as 5cms in height.

Reports of creatures resembling Yowie have varied significantly in estimates of their height. Given that some family groups have been spotted, it would appear that there is a very healthy breeding population, with juveniles and females sometimes living apart.

There have also been multiple sightings of lions, tigers and panthers, although such big cats are not native to Australia. The question is how did they, and the Yowie, get there?

As you can see in Figure 1, there was no land bridge to Asia in the last ice-age. Were some "Others" playing games? The tall hairy Yowie are still reported as present, but where have the others gone? Are they still there somewhere, did humans hunt them to extinction, or did the "Others" move them on? It must be remembered that there is reputed to be an "Other" species, the Urmahs, who are tall felines. They are generally bipedal, but that doesn't mean they have to be if they are pretending to be something else. I haven't heard of a basically canine species which could pretend to be Thylacine.

At sea there have been plenty of sighting of a large sea monster with a long neck, up to 20m long, and also of coastal sightings of metallic objects thought to be like submarines but in water too shallow for submarines.

There have also been several records of another monster, which has been given the scientific name "globster", found dead on a beach. This has been described as having a 10m body and being almost 3m high, hairy, with 6mm skin then a layer of fat then solid meat. The interesting thing was that it appears boneless. Could it be an unknown relative of the giant squid? If you remember, back in Chapter 2, I suggested that a cephalopod might be the best bet for evolving into an intra-terrestrial.

The UAPs which have been reported are typical of those seen worldwide, with some exceptions:

- Large and small cigar shapes
- Rectangular bodies as big as tramcars
- Hexagonal
- Large double disks with portholes between
- Single disks below with a large dome on top
- Disks with a domed cockpit
- Egg shapes
- Spheres of varying size

In addition, there have been tiny metallic devices of various shapes reported as simply falling out of the sky.

Inland from Sidney lie the Blue Mountains. There are claims that there is an UAP base there. Typically, the area has been called the Blue Mountain Triangle.

Certainly, there have been some very unusual happenings there over recent years.[267] There are apparently 600 or so reports of unidentified flying objects that have been seen over, or on, the Blue Mountains. The Jamieson Valley and its surrounding environs has been the scene of many spectacular UAP events. Also, there have been repeated sightings of UAP's that seem to just appear as if out of nowhere in the almost impenetrable forest regions to the south of Mount Solitary and beyond the backwaters of Warragamba Dam.

However, the authors contend that this site contains a massive US/Australian base, bigger than Pine Gap, where the US manufacture their own spacecraft. They make no claim that "Others" are present in the base, but suggest that they may be in contact with them. Certainly there are enough sightings of "Other" craft in the area to suggest that the "Others" are keeping a close eye on them. I leave you to read their book, and to form your own judgement, but this doesn't explain where all the "Others", sighted in these areas, come from:

- Yowie   4
- Tall Human-like   2
- Tall Fair Nordics   4
- 1.8m no neck   2
- 4m lizards   1
- 1.4m red hair   3
- 1.0m dwarf   8
- 0.5m Little People   10

These figures suggest that the smaller "Others" predominate in Australia, along with the Yowie.

A Darlington[268], Perth, man says he's captured pictures of hundreds of UAPs from the veranda of his home. It began when he was taking photos of clouds to test out a new camera and he noticed a "smudge" that, when enlarged and enhanced, "had some structure to it, suggesting it could be some sort of craft in the sky". He says that, since then, he has identified a dozen different UAP types including round, square and saucer-shaped craft, posting the photos to his website wispyclouds.net for extra-terrestrial buffs and sceptics to ponder.

Overall, there are no positive indications of any maritime-based colonies in the area. They are either keeping their heads down, or are land-based.

ix.   Malaysia and Indonesia

The pinwheels seem to be similar to the one sighted in the Gulf of Carpentaria in Australia and the multiple sightings in the Persian Gulf.

Beyond that, there appears to be no doubt that tiny "Others", Little People, about 7.5 cms tall, have set up a colony in Malaya, maybe in Terengganu on the eastern seaboard. There may also be a second species present, perhaps dwarfs, towering over the first at a full 1.0m.

x.   Japan & the Dragon's Triangle

There were very few sightings of "Others", but I have been able to look at those UAPs where there was a description provided:

- Cigar shaped   6
- Disk   5
- Triangular   2
- Rectangle, Sphere, Egg, Cone   1 each

The island of Pohnpei in Micronesia contains the remarkable ruins of Nan Madol[269], which might prove to be the residual evidence of an underwater base. Also, in the mountains of Fukushima Prefecture[270] in Japan near Senganmori Mountain, the locals report frequent sightings of "Others" with wings and hawk-like features. These

Avians sound similar to Anzu of ancient Sumeria or Horus of Egypt, and this may be their long-term Colony

xi. Coastal Russia

Most of the sightings recorded in this table come from three books.[271,272,273] The single sightings do give a pointer to abductions, claims of which seem to become far more common after the 1950s.

There are very few "Other" sightings in this list, with 6 tall, 2 Tall & hairy, two Dwarf-size, and one each of an "Other" with a tail, one with no neck and one winged "Other". It would seem that "Others" aren't too keen on being out in the cold either.

Amongst UAPs there is more of an indication:

- Sphere   14
- Disk   8
- Rectangle (all 12 in one sighting)   12
- Gigantic   6
- Triangle, Cigar, Cylinder, Bell   1 each

xii. Hawaii

All these sightings come from one book[274]. There does seem to be a wide range of "Others" described amongst these.

Menehune is the islander's name for a tiny "Other" less than 1.0m tall, which is known to play with children. In Mexico they are known as Chaneques.

It is clear that there were "Others" in Hawaii before the atom bomb, and that they have an underground base or colony which, based on the sightings, is likely to be in Oahu.

d) ATLANTIC, CARIBBEAN & BALTIC

i. Nova Scotia, Labrador & Newfoundland

These sightings come from 2 books[275,276].

There were only four sighting of "Others" in this area, three of them Little People, but there were several claims of abductions amongst the single-witness testimonies. The Atlantic is very deep between Greenland and Canada, and this was not covered by the North American Ice-Sheet during the last ice-age, so an "Other" base could be hidden there. Sea-ice would have been common, but the ice-sheet itself would not have reached out very far.

However, on the Quebec-Newfoundland border there is a mine, much deeper than others nearby, and which is guarded by the police, where Little People are reported to have been in residence for a long time – a Colony!

ii. Bermuda & the Southern East Coast of the USA

The Bermuda Triangle is not included in this area. It falls within the area covered by Puerto Rico below. Many of the sightings off the east coast of the USA are so far out to sea that their location cannot be specified in terms of their proximity to any particular place on the mainland coast.

There were very few sightings of "Others" in the whole of this area:

- Tall 2.2   1
- Nordics   3
- Large winged   1

- Grays   7
- Dwarfs   1
- Small   1

There were plenty of sightings of UAPs, including the UAP train on its way from Alaska to Brazil reported on elsewhere.

- Triangular   1
- Disk   12
- Bell   1
- Small Square 2
- Cigar   1
- Unspecified   24

These only give the hint that possibly the Small Grays have moved into the area, but no indication of where they have set up home. An unexplained explosion, heard over North Carolina, points to the fact that they probably have.

The closest other possible suspected UAP sites are in Newfoundland, the Bahamas and Puerto Rico, unless there is one within the Bermuda Triangle as claimed in several publications. However, there seems to be little or no evidence to support this.

iii. Gulf of Mexico & Its Coastline

The majority of the sightings recorded here come from three books.[277,278,279]

The underwater topography of the area is shown in Figure 3.

At least there are a few underwater sightings:
- The "Others" sighted by divers working on an oil-rig
- The two submersible UAPs
- The three long-necked sea monsters

Generally, the UAP sightings in the area are typical for North America:

- Disk   18
- Small Disk (1-2m)   7
- Sphere   12
- Egg-shaped   4
- Triangular   2
- Cigar   2
- Tadpole   1

As you might expect, the Small Grays preferred mode of transport tops the list, with spherical UAPs second. I'm never sure whether "Egg-Shaped" refers to the shape of the egg's shell, or as it comes out of the pan when fried. Both are credible. I suppose a "Tadpole" is a disk with a tail at the back.

There is some difference in the frequency of "Others", even though Small Grays still top the list, with so many US states bordering the Gulf:

- Small Gray   8
- 1.5m Tall   3
- Very Tall   2
- Tall & Hairy   2

- Tall   1
- Bat-Like   2
- Little People 2
- Lizards   1

Here the Little People are accused of abduction of young children, to the despair of their parents. However, in Mexico, they are known to look after them, feed them, play with them, and send them home after about 5 days.

They call them Chaneques. If they have been around long enough to pass into folklore, their presence is not modern. Their Colonies are probably underground or in lakes rather than under the Gulf.

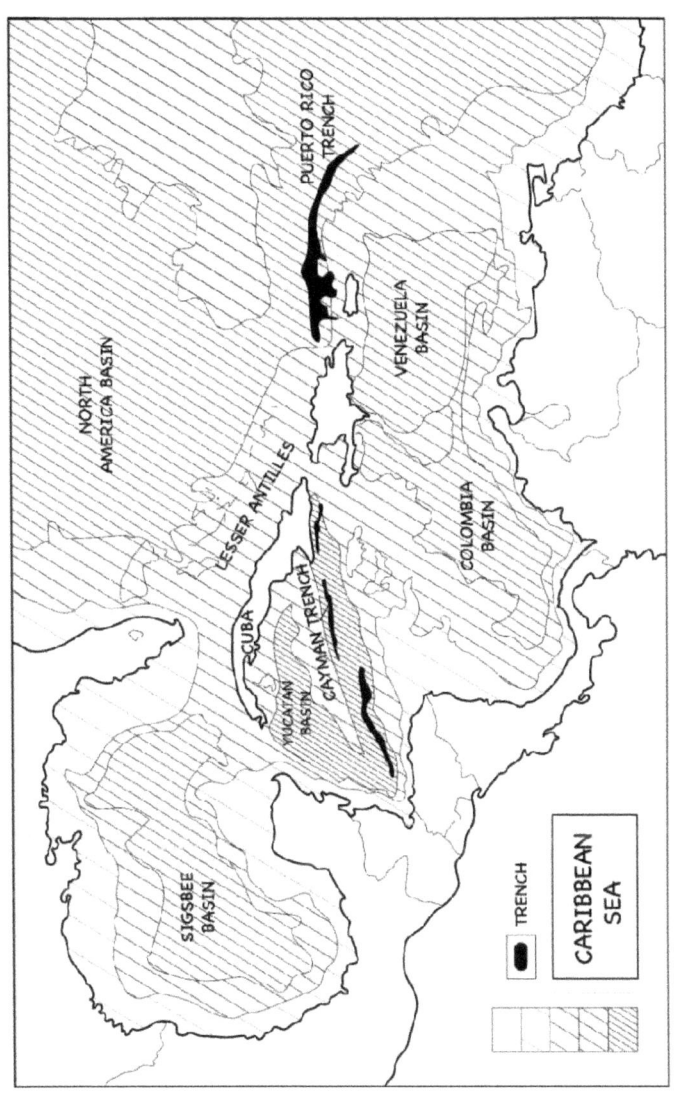

Figure 3 Topography of the seafloor of the Caribbean Sea

iv. Puerto Rico, Bermuda Triangle & South Caribbean

The majority of sightings in this table come from one book[280], together with additions from two others.[281,282] For its size, Puerto Rico is almost overflowing with various types of "Others". I have omitted dozens of sightings of spherical UAPs and of Small Grays. Nevertheless, many species are represented in the sightings:

- \>2m tall thin — 2
- 2m tall hairy — 3
- 2m tall Nordic — 10
- 1.3m Grays with large heads and eyes — 14
- 1.3m – 2m creatures with wings — 6
- 1.3m grays with crest — 1
- Dwarf — 10
- Little people — 10
- Canines — 3
- 1.3m carnivores – Chupacabras — 2

The predominant species appears to be Small Grays, even though they only started arriving in 1978, and there are repeated single-witness reports of their abducting people. They are regularly reported in the vicinity of Lake Cartagena and in the area of San Yungia, in both of which they are suspected of having bases. Both of these seem to be relatively recent, whilst the Colony reputed to be in the nearby Puerto Rico Trench, could be much older.

It is interesting to note that there is a disused gold mine in San Yungia. "Others" are well known for their interest in gold.

The next most frequent sightings are of Nordics, Dwarfs and Little People. The Little People are not known to like being underwater, and the Nordics have been around for longer than the Dwarfs so, on the balance of probabilities, the Nordics could occupy the Colony in the trench.

One of the islands of the Bahamas, Andros Island, hosts the US naval research base AUTEC. It is claimed[283] that there is an "Other" base in the deep water close nearby, linked by power and communications cables.

In Venezuela there were two attempted abductions by "hairy Dwarfs" who proved to be very strong, with apparently unbreakable bodies.

v) <u>Cuba & West Caribbean</u>

The only "Others" sighted are one Bigfoot and several large Winged creatures. It doesn't give any indication as to what the UAPs are doing on the island, although sightings of these could be used to support the assertion that there is an UAP Colony in Guantanamo Bay, even if it is not clear who is in residence.

vi. <u>Brazil</u>

Almost all these records come from one book by Thiago Luiz Tictti.[284] I have extracted records for the maritime provinces of Brazil. He does not claim that the list is complete, and he sometimes comments in a report that there have been other sightings in the same area, but he doesn't list them.

There have been many spectacular sightings in the rest of Brazil, but here I am specifically interested in underwater bases and colonies.

In my area of interest, "Other" sightings have been recorded as:

- Over 2m tall   14
- 1.9 m plus, tall and fair   5
- Human-like   6
- 1.3 m tall   33
- 1m tall   11

One tall reptilian has been reported.

Of these "Others" the group with heights about 1.3m and under predominates. These could be either the Dwarf-like "Others" or Small Grays, but there have been few claims of having seen the distinctive head and eyes of Small Grays.

Some of the reports of failed abductions describe the victim holding onto a tree, resulting in a protracted tug-of-war. This rough handling is more reminiscent of Reptilians rather than Small Grays with their mental abilities, although it would be a mistake to discount the tall fair Nordics who have been known to abduct humans.

There are fewer sightings of "Others" reported along the Brazilian coastline, than along the shorter Argentine coastline, but is much easier to remain unobserved in jungle.

Given the large number of UAPs reported in Brazil as a whole, there must be a number of UAP colonies or bases in the rainforest and in the Andes. There is reported to be one "Other" Colony in the ocean

100 miles south of Rio de Janeiro.[285] Tall "Others" have claimed to own this, but do they share it with any other species?

The sky-train event of 1913 describes a multitude of UAPs travelling line-ahead from at least Saskatchewan down the east coast of the United States, over Bermuda and down to the most easterly point of Brazil where they vanished, possibly into a newly created colony. What species were they? Reports from the surrounding states give no help.

   vii.   Argentina

In the Argentine as a whole, there are reports of at least two "Other" bases or colonies, one south of Cartagena and one off the coast, as well as a wide variety of "Others". Over the country as a whole:

- The majority presence is of tall thin entities well over 2m tall. Also at 2m, is a broader hairy version described as being like a Yeti.
- There is also a Tall entity with a small head so that it appears that its arms come out of its neck.
- There are also sightings of tall lizard people, always underground.
- At typical human height is an orange skinned entity. Also a beautiful blond lady is described who warns of floods, and hovers above water.
- There are reports of 1.5m tall entities with large heads, big eyes and often wearing a light on their head.
- The 1.2m tall creatures are often described as gnomes or dwarfs, and seem to have green skin.

- The second most common species is about 60 cm tall - the Little People

In the following table, I have confined myself to reporting only sightings of "Others" in Argentina's four coastal provinces. All these records come from one source[286], although that one is prone to duplicate records.

The presence of various "Others" here is just as diverse:

- The Little People   45
- Very Tall   37
- Nordics   20
- Dwarfs   20
- Small Grays   19
- Felines   6
- Lizards   5
- Tall Orange Skins   3
- Yeti, No necks, Mothman, Canines 1 each

It can be seen that, inland, there are multiple species of "Others", but on the coast, there are more sightings of very Tall "Others" – Arcturians - and the very small 0.7m Little People. Perhaps they work together. There was some form of a major upset in 1968, when there was a multiplicity of sightings of very Tall "Others", but it is not evident what the cause was.

In addition to any inland bases in Argentina, there is reported to be an underwater colony off-shore at Mar del Plata[287], a few hundred miles south of Buenos Aires. This appears to be the domain of the very Tall "Others". This consistency tends to confirm that coastal

Argentina at least has been colonized by very Tall "Others", perhaps together with the fairy-like Little People.

viii. <u>South Africa</u>

These sightings all come from the one book[288]. Interestingly there are many sightings of sea serpents also recorded, stretching into the 1950s. I have omitted these.

There was a flurry of UAP sightings in the Eastern Cape in 1972. Prior to that, in 1914, a single UAP was reported to have over-flown a lot of the country. Was this reconnaissance prior to establishing a colony in the Eastern Cape? The presence of a group of 20 one meter tall "Others" near Durban in 1986 suggests that the Little People might be in occupancy.

ix. <u>Portugal, the Azores and the Canaries</u>

It is possible to produce a narrative based on these sightings, provided they are accurate:

> *There was an "Other" base at first in Tenerife until one of their tunnels was found there in 1912 and they moved to the mainland. They remained there in the Algarve, until their underground tunnels there were found in 1962, and whoever was in occupancy felt they had to move again. Their base or colony is now back in the Canary Isles but presumably much better hidden.*

However, it would only take one error to change this. If the tunnel found on Tenerife was really a lava tube, we end up with the initial base being in the Algarve. I suppose that the same could be said about the tunnels in the Algarve, but it is evident that something significant happened to sightings in Tenerife in the 1960s, with none recorded before then. Either a base or colony was moved from the Algarve, or a colony was set up from scratch.

Some massive UAPs have been sighted in the area during that period and afterwards, so the colony would have to be pretty well concealed, probably with an underwater entrance because, after these sightings, the local inhabitants would have had plenty of incentive to go looking.

Too many different species of "Others" have been sighted since then to determine which is in residence, or whether it is the classical HMO – House of Multiple Occupancy.

x.   Norway

Records of strange happenings in Norway[289] go back to the 1700s. Many of these describe sea-serpents, lake-monsters, the disappearance of children, stories of strange creatures, and attempts to lure young women away. This is so much part of Norwegian folklore that it cannot be taken as definitive evidence of "Other" activity but, if a colony has been there long enough, it may well become part of traditional fables. Many stories stand out as offering this possibility.

It is difficult to tell whether there is an "Other" colony in Norway, where it might be, and what species of "Other" might occupy it. The best bet, on the basis of the multiple UAP sighting over many years is Hessdalen near Trondheim. This could either be an underwater colony or one in a mountain top. There appear to be multiple sightings of small "Others", particularly at Honefoss, near Oslo, but also riding within UAPs in many other sightings.

Adding to these considerations are the events which have occurred in next-door Sweden:

- In 1946 there were over 50 sightings of 3 - 4m long cigar shaped UAPs which seemed to be deliberately crashing into freshwater lakes
- From there on, there have been multiple sightings of flying saucers of the sort familiar to North Americans.

It would seem that Norway, combined with Sweden, is densely colonised by "Others".

xi.   Sweden

Sweden appears to have a major infestation of "Others.

This commenced long before the Second World War, and may be a prime example of "Other" colonisation of the whole of Scandinavia, with the colonists initially trying not to interfere in our Human society. The Atom Bomb changed this.

However, for the purposes of this book, which is concentrating on "Others" in our waters, the arrival of the cigar-shaped UAPs in the Swedish lakes is my primary interest, along with the possibility of a colony in the Baltic or a Swedish lake,

There are records of the arrival of 6 of these so-called "Ghost Fliers" in 1934, followed by at least 50 in 1946. This is not to say that nothing happened in between, but we only have a few records or sightings during war time. Did the Ghost Fliers continue, or did the war disrupt them? We don't even know what type of "Other" to look for.

Overall, there have been sightings of entities less than 1m tall, beautiful human-like entities that look like teenagers, adult size humans that seem a bit strange, very tall entities and Yeti-like entities.

There have also been reports of strange creatures in lakes, going back years - Sweden's "Lake Monsters". One major example of this is Lake Storsjon, joined to the sea, where the first reported sighting was in 1820. It was described as having a body 8 – 10 metres long, the width of a rowing-boat, with a 10 metre long neck almost as thick as a man's body and with a head like a red-gray horse with a white mane. The last reported sighting was in 1965.

The distribution of sightings across Sweden is fairly uniform, if reckoned in terms of land area, but not if considered in terms of population.

The whole country has 29 sightings, whilst Lapland has 10:

	Country	Lapland
• Disk	16	10
• Sphere	4	
• Cigar	4	
• Oblong	3	
• Pyramid	1	
• Triangle	1	

This gives a total of 29 UAPs, one third of which are in the least inhabited areas. The same statistic applies to disk-shaped UAPs if considered alone. This suggests that something different is going on there, and that we should look for a possible "Other" base.

When we look at "Other" sightings, a different picture arises:

- Dwarf   11
- Small Gray   7
- Cyclops   2
- Slugs   2
- Tall Thin   2
- Little People   2
- Tall Black Bird   1
- Hairy 2m   1

There is only one "Other" sighting in the whole of Lapland.

In practice, it is probably difficult to differentiate between the Dwarfs and the Little People in these sightings, so it may pay to count them as one.

It should be noted that the King of Sweden was visited unannounced by 3 Dwarfs in 1941[290]. It may be that they were declaring their presence and seeking permission to set up home there.

So what we could have is a colony of Dwarfs in northern or southern Sweden, with some Small Grays and a smattering of the other common species. There may well be a separate colony, perhaps with a different species in residence.

Unfortunately, we have no idea what was carried in all those Ghost Fliers. They were Little People sized, but were they land-bound, amphibious or a water-breathing species?

e) INLAND WATERS

i. Russian Lakes & Rivers

These records come from three sources[291,292,293]. There are many recorded observations in the Ukraine which have involved sightings of "Others". There have been some which are probably Small Grays in their flying saucers which are familiar to North Americans. However, the majority have been of spherical UAPs carrying varying types of very tall "Others". There have been 3m humanoids looking like enlarged modern human kind, and it has been suggested that these are descended from the Titans of ancient lore. There have also been similar entities with much smaller heads, and very hairy versions described as like a Yeti.

Some have claimed to have seen webbed feet and hands. However, there is the possibility of these being part of their immersion kit which also includes a spherical helmet as described in the Lake Baikal[294] incident in 1982.

These spherical UAPs are recorded across the whole of Russia/USSR, although there are fewer on the coastlines where the proportion of flying disk increases. This implies that Tall "Others" are also widely spread over inland Asia. It is not clear whether they are showing a preference for fresh water over salt water. Sightings in the Baltic and Black Sea suggest they don't have a total aversion the salt water.

Looking at the large number of single-witness sightings relating to abductions, it would seem that, in Russia, the role of Small Grays in the USA has been taken over by the Tall "Others, undertaking much the same procedures.

Given the multiple sightings in Lake Baikal, there could be a colony there, but they could well have a second, western site somewhere.

ii. The Great Lakes

It is suggested that there are at least two UAP bases in the Great Lakes.

Lake Superior is 1,333 feet deep, Lake Michigan 923 feet, Lake Huron 750 feet, and Lake Ontario 802 feet. These would not be deep enough to hide a base from current detection methods, so any UAP bases would need to be buried beneath the lake bottoms.

There are no sightings of "Other" entities reported on the lakes, so we need to look for any abductions reported in the surrounding areas if we are to identify the species occupying any bases there.

Lake Michigan is the location of one of the claimed Mystery Triangles, and this one includes, along with various UAP sightings, disappearances of both ships and aircraft, with some leaving no trace.

It may be that the Great Lakes are purely an UAP refueling stop. They seem to require water as part of their propulsion system, so they may simply using one of the best sources of fresh water in a stop-over before heading for their Earth base somewhere else, or outwards.

iii. Northern Italy

Mount Musinè[295] in northwest Italy near Bonere has had a reputation for strange goings-on right back into Roman times. There are also nearby petroglyphs that are claimed to show "Others" and UAPs.

There have been fewer sightings there of late, which might suggest that the base may have been moved.

During the early years of the 1800s, there were several reports of an explosion followed by an earthquake in Piedmont, and of various materials such as red soil, coal and yellow rain falling from the sky. This was accompanied by increased UAP activity at Mount Musinè.[296] The explosions seemed to be centred under the Apennine Mountains. It has been said that the explosions caused the earthquakes, and that someone was excavating a new base there. Whatever the excavation technique may have been, the material which was removed could have been ground up and ejected into the atmosphere to fall where it will.

These references almost all come from the one book[297], but it contains many single-witness reports, which I have chosen to discount. There is a more specific listing,[298] but unfortunately, this only catalogues sightings up to 1978. There is reputed to be a second volume continuing up to 2020, but this is currently unavailable. In the meantime, I have added a few extra more modern sightings from another book.[299]

It is interesting to see the distribution of "Other" species in the area:

- Over 2.2m humanoid   5
- 2.2m Bat-like   2
- 2m Lizard upright   2
- 2m no neck   2
- 2.2m hairy   2
- 1.7m Grays Big Head   7
- 1.3m Helmeted Dwarfs   15

- 1.0m   9

It appears that the smaller species of "Others" are in the majority with the Small Grays pushing in. These are also present in many of the single-witness abductions.

Looking at the distribution of "Other" sightings around the provinces, it is not easy to identify the presence of any bases or colonies. There are sightings of disks emerging and entering the Adriatic in the Marches province, but the terrain there does not really stand out as suitable for a possible underground or undersea base. On the other coast the province of Liguria, which has more sightings for its size than might be expected, has predominately orange spheres in its area. Here the terrain, dominated by the Apennines, could suit an underground colony with an underwater exit.

Perhaps there are now two sites in northern Italy, one on each coast.

There is a significant presence of Dwarfs in the area, particularly in Liguria. It is certainly possible that Dwarfs have an underground colony with an underwater entrance near Genoa.

Chapter 4
WHAT "OTHERS" DO WE KNOW ABOUT?

a) <u>GENERAL</u>

As I have already mentioned, there are four possibilities for the origins of "Others":
- A more advanced version of the Human Race (Telosians)
- Species which evolved on earth faster or earlier than humans (ITs)
- Species from parallel universes (PUEs)
- Species from off-Earth (ETs)

The presence of one of these does not preclude the presence of any or all of the rest.

I have been trying to obtain realistic images or descriptions of some of these "Others" but the various publications on the subject are contradictory and, on some occasions, quite fanciful.

It is clear from many of the sighting reports, that it is dangerous to get too close to UAPs. They appear to be either dangerously radio-active, or emitting dangerous levels of Electro-Magnetic radiation, leaving humans with burns on their skin or, at worst, cancer and even death afterwards. It is ironic that many "Other" races are reported to be concerned about our use of atomic power, yet they are cavalier about the effects their technology has on us.

It would be a grave mistake to assume that these colonist "Others" only wish to live in peace with us. It would appear that the Galaxy is a dangerous place, and the various "Other" species are often playing a deadly game of one-upmanship. To some, we are, at present, simply a possible food source, whilst others are trying to protect us from their excesses. Unfortunately, some of these predators are colonists on Earth.

I will deal with the different types of colonists in groups as appropriate. I am going to use the names given to them by the authors of various books on the subject.[300]

My main difficulty is that some of these sightings sound like fantasies, but they have been witnessed by at least 2 people, so I must treat them as credible.

b) WATER MONSTERS

Right the way into the 1950s, there have been reports of monsters of the sea and lakes and, whilst it was fashionable to ridicule such sightings as a fancy of the imagination of early sailors, this accusation cannot be laid at the door of more modern sailors. Sightings from the 1850s onward have included:

- Norwegian lake-monsters, whatever they may be.
- Sea-serpents about 20m or longer. These can be very aggressive. They could have evolved from a creature such as the oarfish.
- Long necked beasts with flippers and a tail. The neck may be typically 5m long. These are apparently like the supposedly mythical Loch Ness Monster.
- Boneless "Globsters" whose bodies have only been found when dead or dying. These could be Cephalopods - an evolved squid of octopus.

These could all be evolved versions of Earthly creatures, and could match or exceed us in intelligence. Remember though, that Cephalopods may not be originally from earth, although their arrival was so long ago that this distinction is rather moot.

They need not necessarily have bases or colonies as we understand them. They would, after all, be living in their natural environment.

None of these could be mistaken for the early water gods which are honoured for teaching our early ancestors fundamental technology. Mostly, those were at least partly human in appearance.

c)   **GIANT APES**

These occur all over the globe, may have local names such as Sasquash or Yowie, and may differ in colour, but they are essentially the same species. Males are typically 2.2 metres tall with smaller females.

They have been observed in small family groups of male, female and child, when they tend to avoid confrontation. Single males can be much more territorial, as can what are probably smaller juveniles.

Humans are reported to have first met these creatures in Tibet, where the received the name Yeti[301]. This is the title I will use rather than any of the ones that have been used elsewhere in the world since then.

If they originated in the mountains of Tibet, they could have reached the Americas in the last ice-age when there was a land-bridge between north-east Asia and Alaska, thousands of years ago.

Now we come to the interesting question: How do they range so far? They have been observed in Australia and New Zealand, and there was no land bridge to Asia during the last ice-age. Also, they have been reported to use weapons to immobilise humans, although this is not common. They are sometimes reported as wearing some type of medallion with technical capabilities.

The proof is not there, as we have no photographic evidence, but it is extremely likely that Yeti are Extra Terrestrials and are far more intelligent than we give them credit for. They live an idyllic life, which may actually be a life-style choice. They also have the intelligence to avoid the human hunters who want the fame for killing one, and have never left a carcass for someone to use as a trophy.

They may have access to the necessary technology to fly around Earth, or perhaps are able to call on help from other ETs.

They live off the land and are not hostile to humans. Our ideal neighbours.

d) THE GRAYS

In practice there are at least three species of Gray "Others",[302] two of whom work together, sometimes doing very nasty things to live human abductees.

- Good Grays

Zeta Humans, from the constellation Cetus, are from 1.0m to 1.5m tall. They have thin bodies, bulbous heads, long arms, three fingers and eyes twice the size of ours. They are described as joyful and intelligent.

- Small Grays

Small Grays, originally from the constellation Reticulum, are from 1.3m to 1.6m tall. They have thin bodies, bulbous heads, long arms, three fingers and very large eyes which wrap around their heads. They practice genetic manipulation and reproductive system surgery on their abductees. They are reputed to have no empathy or remorse for their victims. At first glance, it is difficult to differentiate them from the Good Grays.

- Tall Grays

Tall Grays are also originally from Reticulum, are 2.0m to 2.5m tall and also have bulbous heads, thin bodies, and giant insect eyes which are twice the size of humans.

e) <u>TALL WHITES</u>

There are four types of tall whites which have cropped up in the preceding analysis;

- The Cyclops[303]

These are described as having a single eye and tall at 2m, but only one of the "Others" sightings was that height. The remaining 10 were much shorter. Were they juveniles or are these a separate race? They were described as being well adapted for swimming.

- Reds

These are described as being 2m tall with red or orange hair and orange skin. Nothing is recorded on what they do.

- No Necks[304]

These are again about 2m tall, but they have no apparent neck, so that it looks as if their arms come out of their head. They are occasionally seen moving around but there is little to describe what they are doing.

- Over 2.5m Whites

These are the tallest of the whites which have been recorded here, and are describes as of slender build. They are termed Arcturians in some of the publications,[305] on the subject of extra-terrestrial species. They are generally known as one of the "Good Guys", working to improve mankind's lot. Internationally they appear in sightings about as often as the Small Grays.

In speaking with contactees, the Arcturians have stated that they have colonies in or near the underwater trench which runs up the east side of South America although, considering their prevalence, they must work from other locations as well.

f) NORDICS

This species is the most ambiguous observed. They look almost angelic, are reported to render assistance to humans in trouble, but it is claimed that in Russia they have undertaken abductions in the same manner as Grays, but not as frequently.

However, there is more than one Nordic-looking species. The Cassiopeians[306] are the one Nordic species which has webbed fingers. They have been described as a benevolent species, but we may be mistaken. Perhaps it is they who are at work in Russia, using all the lakes there as homes. They may have a Colony in Lake Baikal[307].

The most likely benevolent Nordic species is the Pleiadians[308], who would certainly act to assist humans where necessary.

They are recorded as having met with President Eisenhower 1954 at Edwards Airforce Base to offer technology in return for nuclear disarmament. Ike refused, because he would not do this unilaterally. They ended by recommending he did not deal with another species called the Small Grays. Unfortunately he did.

They are over 2.0m tall, with fair hair and a light complexion, making them look Scandinavian. Hence the nickname.

It is likely that they have a long-standing Colony in the trench to the north of Haiti, based on the frequency of sightings nearby.

g) DWARFS & LITTLE PEOPLE

There have been sightings of Dwarfs[309] all over the world, and they vary greatly in skin colour and facial features. In general they are between 1.0m and 1.3m in height, and can be very hairy. They are extremely strong and, according to those who have got in a fight with one, virtually indestructible.

Dwarfs have become part of our folklore, which shows that they have been here a very long time. They were openly spoken about in Roman times, as individuals if not as a race.

Green skinned Dwarfs have given rise to the standard term for "Others" – Little Green Men – and the standard name, Martians. There is no evidence to suggest that they actually are Martians.

In the Middle Ages, they tended to fade out of our consciousness, although there were always rumours of hidden Dwarf races in distant lands. "Beyond the Atlas Mountains" was a favourite.

It seems likely that, in modern times, the Dwarfs originally had a colony in Mount Musinè in Northern Italy, and that they have re-located to the Genoa area, perhaps because they could no longer remain hidden where they were.

It is doubtful whether this is their only colony, but their others could be inland in obscure mountains where their secret presence is not under threat.

Their construction technique seems to include very loud explosions, followed by earthquakes. If this happens elsewhere, such as the Catalina Strait or the Eastern United States, this is an indication that they are at work again, even though it may be on behalf of another species.

Smaller than the Dwarfs are the Little People, ranging from about 1.0m tall right down to about 0.15m. Their tiny size has led to their absorption into native folklore around the world, under whatever name they are given locally. Again, they must have been here for a long time.

The Little People are not all one race. In Mexico, where they are known as Chaneques, they are known to steal children to play with them, to look after them for a few days, and then to return them safe and well fed.

There, it is likely that they have at least one colony, claimed to be under a shallow but vast lake near Guadalajara called Lake Chapala. As they have been sighted on both coasts of Mexico, they probably have at least two sites somewhere.

In Mexico's case, the Little People seem relatively harmless, but around the world in Malayasia, things are different. They appear to have arrived in the 1960s and set up home underground, perhaps under Terengganu. These Little People are only 7 centimetres tall and are very aggressive. If threatened, they are quite prepared to fight and to do harm.

In Java, a local man claims to have captured some of these people and exhibits them. He states that, to keep them healthy he has to feed them human blood every month. Apparently, they are not particularly pretty to look at, having protruding eyes – not the traditional fairy image.

It is quite likely that there are Little People colonies all over the world.

h) WINGED CREATURES

There seem to be three different types reported in the previous chapter. There is the black giant that looks like Dracula, Avians, and the dwarf sized sort that looks like an ugly Cherub.

The black giant has been seen worldwide, although it is a loner, and may even have achieved notoriety in the USA as "Mothman"[310]. There is no doubt that this creature scares everyone who sees it, but I have found no record of it hurting a human. It may be different for any animals when it gets hungry. It has been suggested that this creature is a Parallel Universe Entity (PUE), though I do not know how to prove this either way.

The Avians[311] are an ancient race that originally appeared in Sumeria, after the last ice-age working with the Anunnaki. They have hawk-like heads and wings, but it is not clear whether they can fly. There was one there called Anzu who was considered a god. Later in Egypt there was a similar being called Horus who was also considered a god. The only recent record comes from Japan in Fukushima prefecture, where they are believed to have a colony in or near Senganmori Mountain.

The ugly winged cherub has been harder to locate. The one possibility which I have discovered is that there was perhaps a case of partial miss-identification. There is reported to be a species of Dwarf with extremely large ears[312]. These might have been mistaken for wings in the heat of the moment. These Dwarfs are reputed to be able to fly around using their technology, not wings. They have been reported mainly in South America, but not in sufficient numbers to suggest that they have a base or colony there.

i) LIZARD PEOPLE

There has only been the occasional reported sighting of lizard people in my analysis, but they do have a very mixed reputation. They live underground, and one nest was discovered in 1870 underneath Buenos Aires[313].

There are reported to be many species of lizard people, and it has been claimed that one species is a very ancient earth race. Be that as it may, the one we have been seeing in Argentina is a predator on humans, which it abducts. There have been reports on failed abductions where the victim has survived by hanging on firmly to a tree trunk. Lizard abductions lack the finesse of a Gray's abduction because they lack their mental control capabilities. As they seldom leave survivors once they have made their capture, it is difficult to determine the rate of attrition of humans.

j) CATS & DOGS

There are some tales of feline looking creatures turning up in unexpected places, and more tales of canine creatures around the world.

In both cases, there are reported to be "Other" species which could be mistaken for these, particularly if they were deliberately avoiding identification.

Various big cats have been reported in Australia, where they are not thought to be natural fauna. It has been suggested that they are really an "Other" species called the Urmahs[314]. These are naturally bipedal, but there is no reason for them to be so, when they are pretending to be something else.

There are also reports of canines in Australia called Thylacines, known locally, because of their stripes, as Tasmanian Tigers. This was thought to be extinct. There have been reports of "Other" canines in other parts of the world, such as Haiti, but there is too little information to pin them down. Some writers claim that they are called Canis, and come from the Dog Star in the constellation Canis Major, but that does seem a bit too convenient. There was an Egyptian god with the head of a dog, called Anubis. Given that other Egyptian gods are reputed to be Extra-Terrestrial in origin, could it be that this applies to their dog god too? There was an almost identical dog god in Mexico called Xolotl. Both were gods of the underworld.

k) CHUPACABRA

This is another species whose presence is ambiguous.[315] The basic question is whether they are "Others" or escaped "Others" pets. They are also known as Goatsuckers, which gives an idea of their propensities. On the negative side, they are known to attack and kill domestic animals, to drink their blood, and to eat parts. On the positive side they are not known to have attacked humans. They are most common in the Caribbean, Brazil and Florida.

l)   TELOSIANS

In Part 1 of this book[316], I suggested that there was possibly an earlier human civilisation, say 20 thousand years in advance of us, which had already left Earth and now just maintained a few bases here. These are sometimes called Telosians.[317] They look identical to present day humans

m)   A WARNING FOR RESEARCHERS

Different publications give "Others" different names, assert that they come from different star systems, and ascribe them with different attributes. Some authors seem to look on the different species as a creative artistic challenge. It is convenient to be able to have names for them, but it would help if they were consistent. In the meantime, it is hard to know who to believe. The only ones on which there is any agreement are Small Grays, Nordics and very Tall Whites, but there is not enough agreement to decide where they come from.

Despite their presence worldwide, these authors mostly avoid discussing Dwarfs or the Little People. Some have gone so far as to describe them as magical creatures which must be left alone. They do this despite their showing very physical capabilities and responses to aggression.

In the words of Arthur C Clarke, "Any sufficiently advanced technology is indistinguishable from magic".

Chapter 5

WHERE ARE THEY HIDING?

a) <u>BASES v COLONIES</u>

In this book, I have concentrated on the underwater presence of "Others", and especially on those who need to construct facilities to live there.

"Others" have constructed many bases, but not all of these could be considered to be colonies. The particular requirements of a Colony are:

- They must be large enough to enable the inhabitants to continue the way of life they led on their home planet. Sometimes this requirement may be incompatible with an underwater structure where space may be severely limited, and even with subterranean caverns. It is understandable that "Others" seen to roam all over the surface even when they would be better advised not to.

- They must be well hidden, both from the local inhabitants, and from any malevolent "Others" who might visit the planet. Not surprisingly, "Others" exhibit what we would call "Human Nature" and do things for enjoyment even when they would be far better advised to stay concealed.
- They must be self-sufficient so that, if their home planet is destroyed, they can continue to preserve their species. It must be difficult to take such a long-term perspective when, hopefully, "It will never happen". Physical self-sufficiency is one thing, but social isolation could be very different.

In comparison, a Base can be much smaller, only needing to provide the facilities needed for whatever activities are being undertaken there. In addition to the military base(s) needed by the Small Grays to defend this planet, "Others" have bases at mining sites, and perhaps out-stations so that they can keep an eye on their neighbours. Trust is a potentially expensive luxury in this galaxy.

Inevitably, there are going to be more bases than Colonies. One author[318] has estimated that there are almost 50 "Other" bases in the United States alone and, for some reason, this doesn't include Alaska, where we know of at least two major facilities. We only know of two bases in the main part of the US that are Colony-sized amongst the 50 – Dulce and Catalina, although they may be more.

Dulce base is claimed to house about 18,000 "Others", reportedly Lizard People and Small Grays, and some of the activities they are reputed to undertake there would be considered totally unacceptable if we were in any position to prevent them. The Catalina base appears to be more military having, on one occasion, launched 200 disks all at once for reasons unknown. It did sound like a World War Two fighter scramble.

At least, when you are working at one of the smaller remote bases, you have the prospect of going "home" to your Colony,

These suggested motivations, of course, assume that an "Other's" emotions and thought processes are anything like ours. We can understand selfishness, as that is a trait we hold in common with at least some of them, but it is harder to understand altruism or service for the common good.

b) THE COLONIAL ATTITUDE

If you think back to our earthly colonial times, the attitude of the colonists then to the previous inhabitants of the lands they took, left a lot to be desired:

- The Spanish conquistadors stripped South American states of all their gold, and wiped out some tribes.
- The English financed its industrial revolution by robbing India of all its riches.
- The American settlers took the land away from the Native Americans, and even attempted to wipe them out by giving them blankets infected with smallpox.
- The Arabs looked on the Africans as potential slaves, selling them to the Europeans.

In general we looked down on the natives, describing them as uneducated and godless, and believing them to be at best greatly inferior. We could respect their courage at war, but didn't hesitate to kill them in their thousands with our superior weaponry. Massacres of them were considered just, but massacres by them were moral outrages. We even bombed their defenceless villages from the air.

We have to hope that our ET colonists have a more enlightened attitude towards us.

c) THE COLONIES

There are no indications of colonies which precede the last ice-age, although there may well have been a number of "Other" bases then. The Anunnaki would have to have somewhere to work from whilst tweaking our DNA, and that of any more earthly species that they fancied. After the ice-age, the water-based early gods which I've described, the Oannes, Nommo and Dagon, could have been there in Colonies, in the seas, but we don't know where. They could possibly have been in the Mediterranean.

Over all, following the ice-age, there were a number of "Others" (sky-gods) who are described as descending from the heavens, and taking up residence for a period to teach the locals. There are many impressive ruins from that period.

The first arrivals in more modern times were probably the Dwarfs and the Little People, who have been here so long that they have entered into our mythology. This is what makes it so hard for us to accept that they are truely "Others"

- The Little People likely have a Colony under Lake Chapala in Mexico, having survived a crash there. They probably have a second Colony in Mexico by now. There is another Colony or Base in Malaysia, one in Sumatra and one in South Africa. In practice, as the Little People sightings are so widely spread, they may have set up Colonies in a number of places. Their small stature makes them easy to hide.
- Dwarfs appear widely in European folklore, and are represented in several nations' pantheons of gods. They are traditionally famed for their blacksmithing skills. In modern times, there have been fewer reports of Dwarfs, although they do seem to have set up a Colony in Northern Italy near Genoa, with an underwater entrance, and perhaps a Base on the other coast near Marches. They are reported to have set up home in Sweden, perhaps on the Baltic coast.
- The Small Grays first set up a Colony in Sweden but, following the atom bomb and their "Treaty" with the USA, they have also set up at least two Bases there, large enough to be called Colonies. They have expanded down into the Caribbean and across into Europe.
- The web-fingered Nordics have a Colony in Lake Baikal in Russia, and have spread widely from lake to lake as far as the Baltic and the Black Sea in the west and the Sea of Okhorsk in the east.
- If the other Nordics have a Colony, it is most probably in the underwater trench near Costa Rica.
- The very Tall Whites, the Arcturians, have a Colony on the east coast of South America, perhaps in one of their two underwater

bases near Mar del Plata and Rio de Janeiro in the deep trench there. A new base further north was set up in 1913, and they may be there too.

There are two Bases which may be somewhat different. In Hawaii and the Canary Isles, these could well be multiple-occupancy, given the large variety of "Others" species sighted in such a confined area. Could they be some form of oceanic regulator?

Chapter 6

CONCLUSIONS

I have shown that there is every probability that there are at least nine "Other" Colonies present on Earth near the coast, with four of these having underwater access.

The difficulty I've had in identifying Colonies is that we cannot see underwater structures. I have been able to identify concentrations of particular species of "Others", and to hypothesise on the location of their Colony, but cannot prove it.

However, often it has proved possible to demonstrate that the Colony is close to the coast, even if it is on land. As many of the Colonies are fairly new, this might suggest that creating an underwater Colony might be the ultimate objective, but a subterranean one is quicker to produce, and does provide an interim solution.

Some "Others" may never wish to go underground, yet alone underwater. Any Avian might find it too claustrophobic, and prefer an open mountaintop. On the other hand, any remaining Telosians have themselves well-established in mountain and Antarctic caverns, and in underwater structures.

Ultimately, the world's armed forces may well have detailed knowledge of the whereabouts of more Colonies, but governments are still under the misguided belief that they must not tell us mere mortals. Most probably, it suits the Small Grays to leave things the way they are, and our governments are doing what they're told.

Based on this work, UAP investigators will need to further address those questions that I have only begun to tackle. Who are our colonists and where are their colonies?

There will be occasions when the Colonists, in their drive for self-sufficiency, want the same resources that we do. There are certainly mining bases out there already, although we have yet to harvest the sea-bed for all its minerals, as they are capable of doing. At present they are simply helping themselves, but what happens when, inevitably, conflict arises?

Then we need to consider the most vital one: Can we evict those "Others" that are acting against our best interests?

As a world, we are facing many challenges in dealing with the "Others" whether they are just visiting or colonising our planet. Are our governments dealing with them in the best interests of the population as a whole, or simply in the interests of our governing elite?

We need to find out – and quickly!

APPENDIX
Tables of sightings

INDIAN OCEAN
i	Persian Gulf	162
ii	India	162

PACIFIC OCEAN
i	Eastern Russia & The Sea of Okhorsk	163
ii	Coastal British Columbia, Washington, Oregon & Southern Alaska	164
iii	Coastal California,	166
iv	West Coast Of Mexico & Central America	168
v	Peru	169
vi	Chile	170
vii	New Zealand	172
viii	Australia & New Guinea	174
ix	Malaysia & Indonesia	177
x	Japan & The Dragon's Triangle	178
xi	Coastal Russia	180
xii	Hawaii	181

ATLANTIC, CARRIBEAN & THE BALTIC
i	Nova Scotia, Labrador & Newfoundland	182
ii	Bermuda & Southern East Coast of USA	183
iii	Gulf of Mexico & its Coastline	185
iv	Puerto Rico, Bermuda Triangle & South Caribbean	187
v	Cuba & West Caribbean	188
vi	Brazil	189
vii	Argentina	191
viii	South Africa	194
ix	Portugal, the Azores and Canaries	195
x	Norway	197
xi	Sweden	198

INLAND WATERS
i	Russian Lakes & Rivers	199
ii	The Great Lakes	200
iii	Northern Italy	201

PERSIAN GULF & RED SEA

Date	Place	Credibility	Details
8th May 1879	Persian Gulf	Solid	Pinwheel
May 1880	Persian Gulf	Strong	Pinwheel
5th Jun 1880	Arabian Sea	Solid	Pinwheel
4th Apr 1901	Persian Gulf	Moderate	Pinwheel
26th Aug 1902	Nr Gulf of Aden	Solid	Pinwheel
1906	Gulf of Oman	Moderate	Pinwheel
17th Feb 1912	Off Pakistan	Solid	Pinwheel
25th Aug 1925	Arabian Sea	Moderate	Pinwheel
1943	Persian Gulf	Moderate	Pinwheel
11th Nov 1949	Strait of Hormuz	Strong	Pinwheel
7th Apr 1951	Gulf of Aden	Strong	Pinwheel
30th Nov 1951	Persian Gulf	Solid	Pinwheel
2nd Apr 1956	Persian Gulf	Moderate	Pinwheel

INDIA

Date	Place	Credibility	Details
5th Jun 1880	Malabar		Pinwheel
1887	Sikkim		Yeti Tracks
Dec 1914	Darjeeling		"Other" 1.3m hairy
1925	Sikkim		"Other" 1.3m hairy
Sep 1938	Lower Bengal		"Other" giant hairy
Feb 1942	Sikkim		2 "Others" giant hairy
1944	Kashmir		"Other" giant hairy
15th Mar 1951	New Delhi		UAP Cigar
15th Sep 1954	Bihar		UAP 4m saucer
2nd Nov 1954	Calcutta		UAP Bright disk
27th Nov 1954	Madras		UAO Boat flying
1954	Madras		UAP Disk
Late 1958	India		4"Others" 1.0m killed boy
16th May 1974	Bombay		UAP large saucer
24th May 1975	Arunachal Pradesh		2"Others" tall hairy caught
Summer 1977	Bombay		UAP gigantic disk
April 1982	Mysore		UAP 10m cigar 3 "Others" 1.7m gathering samples
Jan 1987	Kashmir		"Other" 1.3m hairy
20 Nov 1992	Calcutta		UAP Sphere
1st Mar 1997	Mumbai		"Other" giant bird
16th Aug 1996	Uttar Pradesh		"Other" like werewolf

EAST RUSSIA AND SEA OF OKHORSK

Date	Place	Credibility	Details
11th Jul 1908	Primorsky Krai		"Other" Tall, winged
22nd Nov 1908	Sea of Okhorsk	Moderate	Light underwater
Aug 1927	Khabarovsk Krai		UAP disk
Early 1943	Bering Sea		UAP running beside ship
Summer 1945	Adak Island, Alaska	Solid	UAP emerged
1951	Sea of Okhorsk	Moderate	Sonar detection of submerged UAP
Jan 1952	Shemya Island, Aleutian Islands	Solid	UAP entered sea
Sep 1956	Blagodatnoye Lake		UAP dish crashed
Jan 1957	Okhorsk		"Other" Tall woman
1957	Bering Sea	Solid	UAP out and in
1970	Kamchatka Lake		UAP emerged
1974	Kamchatka		UAP small sphere
14th Aug 1975	Kuril Islands		Abduction 2 weeks
1977	Sea of Okhorsk		UAP disk
24th Aug 1978	Khabarovsk Krai		Crashed UAP 6m mushroom
May 1982	Kamchatka		UAP 200m cigar
Summer 1982	Primorsky Krai		UAP disk
31th Oct 1987	Vladivostok		UAP huge disk Shot down
Spring 1989	Sea of Okhorsk		UAP egg-shape into sea
20th Jun 1989	Primorsky Krai		"Other" Silvery gray
18th Jul 1989	Primorsky Krai		"Other" 2.0m
6th Sep 1989	Russkiy Island		UAP Rectangular
End 1989	Tikhaya Bay, Vladivostok		Triangular UAP
19th Jan 1990	Khanka Lake		UAP disk with windows
27th Mar 1990	Kuril Islands		UAP 1.3m sphere
31st Mar 1990	Primorsky Krai		UAP Oval
Apr 1990	Uyar		UAP
Summer 1990	Chukotka		UAP cigar & groups
Summer 1990	Primorsky Krai		"Other" 2.0m
Summer 1990	Vladivostok		"Other" Attempted Abduct
1990	Kamchatka		"Others" Tall spacesuits
1990	Timofeevka		"Others" 2.5m
14th Sep 1991	Vladivostok		UAP disk abduction
Nov 1991	Primorsky Krai		"Other" winged man
5th Sep 2006	Vladivostok		3 UAPs Spheres

THE UFO, E.T, ALIEN TRILOGY

BRITISH COLUMBIA COAST & SOUTHERN ALASKA

Date	Place	Credibility	Details
31st May 1831	Puget Sound WA		UAP luminous
2nd Jul 1893	Puget Sound WA		UAP 45m oval
Nov 1896	Vancouver BC		UAPs Luminous disks
1st Feb 1908	Puget Sound WA		UAP Brilliant Red
2nd Feb 1908	Puget Sound WA		UAP Red Cigar
14th Jan 1945	Oyster River BC		UAP silvery cylinder
5th May 1947	Seattle WA		UAP Silver object
18th Jun 1947	Eugene OR		2 UAPs shiny round
4th Jul 1947	Seattle WA		UAP Photo Dot
4th Jul 1947	Portland OR		3 UAPs disks
5th Jul 1947	Seattle WA		2 UAPs football
6th Aug 1947	Douglas OR		UAP 10m sphere
12th Nov 1947	Curry, OR		2 UAPs balls of fire
30th Oct 1948	Grays Harbour WA		20 UAPs eggs
24th May 1949	Curry OR		UAP circular
Sept 1952	Okanagan Lake AK	Solid	Surfaced UAP, taking off
16th Dec 1958	Hawk Inlet Juneau AK		UAP on surface
6th Mar 1960	Fraser River AK	Certain	Large UAP in & out
9th Mar 1960	Strait of Jan de Fuca WA		Flaming UAP into sea
Jun 1961	Seattle WA		UAP Cylinder
Jul 1963	North Portland OR		UAP 10m cylinder
12th Jan 1965	Tillamook OR		UAP Triangle Into sea
25th Mar 1966	Saceanich Inlet AK	rtain	2 UAP Triangles
3rd Apr 1966	Seattle WA		UAP large glowing
Apr 1966	Lincoln OR		"Others" Cyclops
Sep 1966	Lincoln OR		UAP ball of light
13thFeb 1967	Coos OR		UAP ball of light
Apr 1968	Vancouver Island BC		"Other" 1.5m
Spring 1968	West Vancouver BC		"Other" small glowing
Aug 1968	Vancouver BC		UAP 3m disk
10th Jun 1969	Portland OR		UAPs line of disks
Aug 1969	Douglas OR		UAP disk 1.0m
1st Jan 1970	Duncan BC		UAP 15m disk
May 1971	Portland OR		"Other" 3m long arms
Oct 1973	Duncan BC		UAP 25m disk
25th Jun 1974	Vashon Island WA		UAP 10m cylinder
End 1975	Lincoln OR		"Others" taking livestock
9th Mar 1977	Victoria BC		UAP Oval, light on top
5th Dec 1978	Ketchikan AK		UAP Large rectangle
10th Sep 1979	Douglas OR		UAP 10m sphere
25th Nov 1979	Grays Harbour WA		UAP Cigar

BRITISH COLUMBIA COAST & SOUTHERN ALASKA (Continued)

Date	Place	Credibility	Details
10th Dec 1979	Grays Harbour WA		UAP fireball crashed
7th Aug 1980	Tillamook OR		UAP rectangle
5th Oct 1981	Victoria BC		2 "Other" Men in Black
Oct 1981	Vancouver Island BC		UAP Photo high up
31st Aug 1987	Bangor Sub Base WA		UAP Disk with dome
15th Feb 1988	Rivers Inlet BC		UAP exploded
Spring 1995	Tillamook OR		"Others" Grays, orange eyes Mantis
15th Sep 1995	Wasilla Anchorage AK		2 UAPs orange lights
17th Nov 1996	East Vancouver BC		UAP Large black square
31st Mar 1997	Juneau AL		UAP coloured light
25th Apr 1997	Portland OR		"Other" Chupacabra
15th Mar 1998	Anchorage AL		2 UAPs conical
15th Sep 1998	Auke Bay Anchorage		UAP lights
7th Dec 1998	Anchorage AK		UAP lights
8th May 1999	Nikiski nr Anchorage		UAP light
25th May 2006	Ketchikan AK		30 oval UAPs
13th Sep 2010	Ketchikan AK		Spherical UAP
14th Sep 2010	Ketchikan AK		Multiple UAPs
28th Apr 2012	Ketchikan AK		UAP 3 lights
4th Sep 2012	Ketchikan AK		UAP above
8th Nov 2012	Ketchikan AK		UAP above
7th Sep 2013	Ketchikan AK		UAP plasma ball
5th Dec 2013	Ketchikan AK		UAP over airport
7th Aug 2014	Ketchikan AK		Oval UAP
19th Aug 2014	Ketchikan AK		Disk UAP
30th Dec 2014	Ketchikan AK		UAP above
11th Mar 2016	Ketchikan AK		Orange & Red UAPs

THE UFO, E.T, ALIEN TRILOGY

COASTAL CALIFORNIA

Date	Place	Credibility	Details
24th Feb 1904	USS Supply		3 UAPs Large disks
13th Sep 1930	Humbold		UAP fireball into sea
1935	Los Angeles		UAP & "Others"
Summer 1941	Ventura		Small silvery UAP
25th Feb 1942		The Battle of Los Angeles	
Nov 1943	Long Beach		UAP Glowing orange disk
July 1945	Los Angelrs		UAPs Darts of light.
21st Jun 1947	Maury I		5 UAP Debris
2nd Jul 1947	San Francisco		6 UAPs into sea. Footballs
3rd Jul 1947	San Diego		3 UAPs Disks
7th Jul 1947	Mendocino		UAP disk 2.0m
8th Jul 1947	Avalon		6 UAPs Flying Discs
9th Jul 1947	Santa Catalina Island		6 UAPs disks
Aug 1947	Los Angeles		UAP silver green sphere
20th Sep 1947	San Diego		UAP flaming, manoeuvred
May 1951	Solano		Multiple blasts
4th Jan 1952	San Diego		Several blasts
Jul 1952	Culver City		UAP Egg shaped
Mar 1953	Los Angeles		2 "Others" Tall, pricked ears
Nov 1953	Los Angeles		UAP Big silver ball
Nov 1953	Santa Monica		Squadron UAPs
29th Jan 1954	Santa Ana		UAP round with flat bottom
22nd Apr 1954	Saint Nicholas Island		UAP 2m cigar
11th Jul 1955	Santa Catalina Channel		UAP Saucer
22nd Jul 1955	Santa Maria	Weak	UAP Object Emerged
8th Aug 1955	Long Beach		UAP sphere in/out of sea
18th Jan 1956	Redondo		UAP Object Entered
9th Feb 1956	Redondo		UAP Sphere entered
20th Jul 1956	Los Angeles County		2 "Others" 2.2m blond
6th Nov 1957	Playa Del Rey		UAP Egg & 2 "Others"
14th Mar 1958	Sonora County		UAP 1m round
1958	Los Angeles		UAP Large Disk
19th May 1961	Long Beach		12 UAPs 2 explosions
28th Jul 1962	SE Catalina I	Strong	Object Surfaced
1964	Ventura	Strong	Object Emerged
5th Feb 1964	Humbold Bay		UAP under sea
1st May 1964	Santa Ana		UAP large domed
13th May 1964	Solano		UAP sphere
13th Sep 1964	Humbold		"Other" Big & Hairy
Nov 1964	Ventura		UAP from reservoir
2nd Dec 1965	San Pedro		UAP orange into sea
20th Aug 1966	Sonoma		UAP 15m high
Nov 1966	Palos Verdes		UAP triangle
28th Dec 1966	Santa Ana		UAP circular 6m

COASTAL CALIFORNIA (Continued)

Date	Place	Credibility	Details
1st Apr 1967	Sonoma		UAP domed
June 1967	Santa Barbara		7 "Others" 1m tall
Oct 1967	Catalina Strait	Solid	Object Entered
11 Oct 1967	Santa Ana		2 MIB
Spring 1968	San Diego		Explosion Red UAP
Oct 1968	Near Catalina I	Strong	UAP Emerged /Entered
4th Jan 1969	San Diego Mission Bay		UAP Egg shaped
24th Oct 1973	Los Angeles		UAP large round
Oct 1973	Santa Cruz		UAP hovered
12th Nov 1973	Los Angeles		UAP 10m ellipse
16th Nov 1973	San Diego		UAP 6m diameter 3m high
13th Dec 1973	San Diego		UAP 9m long 4m high
17th Jul 1974	Los Angeles County		UAP Cylinder
15th Oct 1974	San Diego		UAP round
Nov 1974	Los Angeles County		2 "Others" Tall &Hairy
Jun 1975	Santa Barbara		UAP 12m disk
Nov 1975	Los Angeles		UAP 0.6m sphere
Aug 1976	Bel Air		Several "Others" gorillas
16th Oct 1976	Ventura		UAP 3m oval
6th May 1977	San Diego		UAP Domed disk
25th July 1977	San Francisco		UAP conical
30th Aug 1977	Sonoma		UAP lights
4th Feb 1978	Orange County		UAP Oval
Summer 1978	Los Angeles County		UAP Oval 25-30m diameter
Nov 1978	San Diego		UAP Disk
15th Jun 1980	Santa Cruz		UAP 1.2m long cylinder
28th Nov 1981	Los Angeles County		UAP
27th Jan 1984	San Diego		5 UAPs lights in formation
Dec 1984	Santa Monica		UAP Ring of lights
9th Feb 1985	Los Angeles		2 UAPs orange spheres
May 1987	Santa Ana		UAP Silver ball
Sep 1987	San Luis Obispo Airport		UAP Intense light, radar
Oct 1989	Santa Rosa		UAP Light cluster
19th Sep 1991	Near Avalon		Multiple. NASA saw
14th Jun 1992	Catalina Strait	Certain	200+ Objects Emerged
30th Aug 1992	Malibu		Abduction
Feb 1993	Los Angeles		2 "Others" 1.3m
13th Mar 1995	Lake Takoe		UAP Triangle
Summer 1966	Beverley Hills		"Other" flying unassisted
8th Oct 1996	Ventura		"Other" Chupacubra
14th Nov 2004	Nr San Diego	Certain	Nimitz Incident
2005-06	San Francisco		UAP 2 triangles. 1 Disk
16th Feb 2008	Miramar		UAP Pulsing orb
22nd Jun 2009	Emerald Bay, Catalina		2 UAPs orb and submerged
14th Jan 2012	Orange County		UAP Orb
Jul 2019	USS Russell		UAP Triangles

WEST COAST OF MEXICO & CENTRAL AMERICA

Date	Place	Credibility	Details
30th Oct 1923	Costa Rica		Sea Monster long Neck
1941	Sonora Mexico		UAP Crash (before Roswell)
1944	Sonora Mexico		Dead "Other" short Large head
1945	Guatemala		3 "Others" 0.5m
1950	Honduras		"Other" Large Bat
1950	Panama		UAP Disk with dome
5th Mar 1950	Guerrero, Mexico		UAP Disk sighted by pilots
10th Mar 1950	Guerrero, Mexico		UAP Disk
May 1952	Honduras		UAP Disk "Other" Small Gray
9th Oct 1953	Jalisco Mexico		UAP Circular Small "Others"
Nov 1954	Costa Rica		UAP Globe
1958	Honduras		Round UAP "Other" <1.0m
30th Jul 1963	Sonora Mexico		3 UAPs huge white spheres
1965	Panama		UAP Large disk. Orange "Other"
30th Mar 1967	Guerrero Mexico		Cigar UAP crashed
1968	Honduras		Giant "Other" 5.0m
1973	Guatemala		2 "Others" Tall
1974	Jalisco Mexico		100m diameter UAP with fin
1974	Panama		"Other" <1.0m
22nd Dec 1975	Sonora Mexico		"Other" 1.3m
28th Oct 1976	Jalisco Mexico		200m diameter UAP
22nd Dec 1977	Guerrero Mexico		UAP 3x diameter of moon
17th Feb 1978	Jalisco Mexico		UAP Egg
18th Jul 1978	Jalisco Mexico		3 "Others" Amorphous
26th Nov 1978	Jalisco Mexico		2 UAPs luminous
1979	Costa Rica		UAP Circular
16th Jan 1979	Panama		UAP Oval
1st Jun 1983	Michoacan Mexico		UAP gigantic sphere
1983	Nicaragua		14 "Others" Tall
6th Feb 1984	Jalisco Mexico		5 UAPs Lights
1985	Jalisco Mexico		"Other" 1.2m normal head
22nd Mar 1986	Jalisco Mexico		"Other" 2.1m
25th Oct 1986	Costa Rica		UAP Underwater
22nd Nov 1986	Jalisco Mexico		"Other" golden figure
Dec 1986	Michoacan Mexico		"Other" tall thin blond
1991	Costa Rica		18 "Others" 1.7m
Apr 1993	Guerrero Mexico		UAP Disk
17th Aug 1995	Sonora Mexico		3 UAPs Orange "Others"
Mar 1966	Guatemala		"Other" Large Bat
1st Apr 1996	Chiapas Mexico		"Other" 0.5m killed sheep
1996	Honduras		UAP Box shaped
1997	Honduras		"Other" 1.3m Pointed hat
17th May 1997	Baja California Mexic		3 "Others" little men
24th Dec 1997	Costa Rica		12 UAPs disks
1999	Panama		"Other" 1.3m wrestled cattle

THE UFO, E.T, ALIEN TRILOGY

PERU

Date	Place	Credibility	Details
Oct 1907	Solimes River		"Other" long necked monster
16th Mar 1950	Lima		UAP glowing disk
19th Jul 1951	Iberia		UAP cigar
5th Aug 1952	Plane Lima to Huacho		3 UAPs saucer-shaped
May 1954	Huamanga		"Other" 1.5m slender woman
30th Jan 1958	Arequipa		UAP saucer
2nd Feb 1958	Arequipa		UAP mushroom shaped
12th Apr 1960	Ancash		2 "Others" Tall
15th May 1960	Cordillera		2 "Others" Tall cured boy
1961	Catalina		UAP row of windows
2nd Aug 1965	Lima		"Other" Dwarf
20th Aug 1965	Cuzco		2 "Others" small from 1.5m disk
20th Aug 1965	Arequipa		"Other" 0.8m shrub entity
25th Aug 1965	Callap		UAP Saucer
3rd Sep 1965	Puno near Lake Titicaca		7 "Others" 1.0m Cyclops
12th Sep 1965	Huancavelica		2 "Others" 0.8m
17th Sep 1965	Huancavelica		UAPs 1 large 4 small disks
29th Sep 1965	Arequipa		"Other" 1.0m Cyclops
1st Nov 1965	Huanuco		UAP Ringed globe
5th Jul 1966	Lima		"Other" 1.5m
30th Dec 1966	Coast of Peru		UAP escorts aircraft
26th Mar 1967	Yungay		2 UAPs
2nd Jun 1968	Lima		6 UAPs soup plates
28th Nov 1968	Lima		"Other" short. Egg shaped head
3rd Feb 1972	Nazca		UAP cigar 15m, 2 Tall "Others"
11th Apr 1972	Cuzco, Lake Huaypo		UAP disk entered lake
Spring 1975	Lima		"Other" 2.2m tall
11th Nov 1975	Loreto		4 "Others" 0.8m UAP 4m oval
May 1977	Lima		0.5m "Other" UAP disk
18th Nov 1977	Arequipa, Yanyarina Sea		3 UAP disks entered sea
18th Nov 1977	Arequipa		3 giant "Others"
25th Dec 1978	San Martin		6 UAPs rhomboid
11th Feb 1980	Puno near Lake Titicaca		UAP disk
Apr 1981	El Collao		2 "Others" 1.2m Cyclops
1981	Paruro		"Others" 2.2m copper hair
1991	Lima		3 UAPs 1x100m 2x50m
1993	Puno		2 very tall "Others"
22nd Aug 1996	Ancash		3 UAPs Tall "Others" get water
Mar 1977	La Libertad		Huge UAP crash
20th Jul 1997	Corsica		UAP Cylinder 7m long
23 May 1999	Recuay		5 UAPs 30m
1999	Lima		"Other" small green head

CHILE

Date	Place	Credibility	Details
5th Nov 1883	Chile		UAP Large disk
13th Apr 1905	Magallanes		Long neck sea monster
Jan 1914	Malleco, Araucania		"Other" 2.0m angelic
Jul 1931	Los Lagos		UAP 50m on surface
Mar 1937	Antofagasta		UAP with lizard "Others"
1945	Antofagasta		UAP with small "Others"
10th Jul 1947	Santiago		UAPs oval
23rd Feb 1949	Chilean Antarctic		UAPs
23rd Feb 1950	Chilean Antarctic		UAPs
11th Mar 1950	Magallanes		UAP Sphere
1955	O'Higgins		Tiny "Other"
28th Oct 1958	Valparaiso		UAP
1958	Tarapaca		"Other" 3.0m thin
1958	Los Lagos		UAP Sphere small "Others"
9th Aug 1959	Magallanes		UAP 2m egg
Jan 1959	Los Lagos		3 "Others" 2.0m
Oct 1963	Malleco, Araucania		"Other" shape shifting woman
1963	Arica		UAP cigar
Nov 1964	Santiago		UAP disk very small "Others"
20th Jul 1965	Atacama		UAP mushroom-like
1964	Chiloe Isle, Largos		"Other" very tall in water
1964	Aracuo, Biobio		"Other" tall long blond hair
3rd Jul 1965	Chilean Antarctic		UAP lens shaped
4th Aug 1965	Santiago		UAP Oval landed
6th Sep 1965	Arica		UAP near aircraft
10th Sep 1965	Santiago		3 UAPs
11th Sep 1965	Villa Alemana, Valpariso		2 UAPs disks
21st May 1966	Colquinto		3 UAPs Spherical
1st Jun 1966	Bio-Bio		3 UAPs
Jun 1966	Los Morros, Santiago		UAP like nut
Jun 1966	Colquimto		2 "Others" 1.2m
4th Dec 1967	Santiago		"Other" Cat-like
May-Jun 1968	Antofagasta many times		UAP 3 "Ohers"
Jun 1968	Cordillera		Procession of "Others" in white
Jul 1968	Santiago		"Other" 0.7m
Jul 1968	Biobio		2 "Others" very tall
Aug 1968	Biobio		UAP disk
Sep 1968	Biobio		"Other" man with wings
27th Dec 1968	Santiago		UAP disk "Others" 0.8m
13th Feb 1969	Valdiva, Largos		UAP disk 3 "Others" 2.0m
10th Jun 1969	Santiago		UAP tube
22 May 1970	Chanaral, Atacana		UAP buzzed ship
24th Sep 1971	Antofagasta		Large Oval UAP
1972	Los Lagos		Large UAP in and out of water
21st Jun 1973	Santiago		"Other" bird man

CHILE (Continued)

Date	Place	Credibility	Details
11th Sep 1973	Araucania		Several UAPs
Jan 1974	Antofagasta		"Other" 2.0m
Summer 1975	O'Higgins		2 tall "Others" collecting fruit
3rd Aug 1975	Verdades ?		UAP bright sphere
1976	Los Lagos		UAP up and down in sea
25th Apr 1977	Arica		UAP and abduction
24th Aug 1978	Colquimto, at sea		UAP sphere. Web-fingered "Other"
29th Aug 1978	Huasco, Valpariso		3 "Others" 2.2m UAP Disk
26th Sep 1978	Santiago		5 "Others" short
16th Dec 1978	Antofagasta		Jet tries to intercept UAP
Nov 1979	Punta De Damas, Santiago		UAP Disk "Other" tall hairy
14th Jun 1980	Santiago		UAP sphere near aircraft
Jun 1980	Cordillera de los Andes ?		UAP 12m cigar. "Other" no neck
28th Apr 1983	BioBio		UAP over football match
1983 or 84	Antofagasta		2 Nordic "Others"
1983	Valparaiso		2 "Others" 2.5m
15th Sep 1985	Arica		2x 0.25m long strange creatures
Apr 1986	Atacama desert		80m UAP disc. Soldiers killed
1986	Valparaiso		2 "Others" 2.5m
Feb 1987	Elqui, Colquimto		"Other" large winged
31st Dec 1990	Santiago		Very tall "Other"
6th Nov 1993	Santiago		"Other" large winged
Oct 1994	Santiago		"Other" large winged
Summer 1996	Antofagasta		"Other" 3m thin long arms
16th Feb 1997	O'Higgins		"Other" 1.5m large head
31st Jan 1998	Colchagua, O'Higgins		UAP sphere blew up on ground
16th Jul 1998	Santiago		"Other" 0.5m hair on head
Aug 1998	Santiago		"Other" 1.9m slim
7th Oct 1998	Valle de Equi ?		UAP 15m hit mountain
23rd Jan 1999	Cautin, Araucania		UAP massive oval mothership
Jul 1999	Coquimbo		2 "Others" 2.0m Nordics
15th Sep 1999	Antofagasta		"Other" 1m Chupacabra
1999	Colquimto		"Others" small humanoids
2014	Tarapaca		UAP

NEW ZEALAND

Date	Place	Credibility	Details
1861	Lake Heron		Giant otter-like animal
Oct 1875	Waitahanui		3 "Others" 1m
1878	Waiki		"Other" big & hairy
May 1883	Masterton		"Other"
1883	Fortune Bay		"Other" Giant 20m Turtle
4th May 1888	Unknown		Oval UAP
14th Jul 1890	East Cape & Portland Light		"Other" 30m monster
May 1896	Wellington		"Other" White
23rd Jul – 9th Aug 1909	All over New Zealand		UAP Airship with gondola & 3 "Others"
1st Sep 1909	Kelso		UAP airship
1910	Coromandel		"Other" big & hairy
16th Jul 1920	Aukland		"Other" White
1919	Canterbury		UAP Ellipse
Aug 1944	Christchurch		UAP Saucer 3"Others" 1.1m
Feb 1952	Coromandel		"Other" big & hairy
6th Jan 1953	Auckland		UAP Cigar
8th Jan 1953	Mosgeil		UAP Triangle
2nd Jun 1953	Havelock		UAP 25m Cigar
19th Mar 1954	Nelson Bay		UAP Disk with dome
Nov 1954	Hamilton		"Other" 2.2m no neck green
6th Feb 1955	Graymouth		UAP Silvery Cigar
13th Nov 1955	Bluff		Cone & Cube "UAPs" surfaced
19th Jul 1956	Gisborne		Spherical UAP
Sep 1956	Putanuru		UAP 2m sphere
15th Jan 1957	Waiki & Balfour		UAP bright sphere or oblong
7th Jun 1958	Mount Egmont		UAP White oval
11th Apr 1959	Invercargill		UAP Sphere
3rd Nov 1959	Woodlands		UAP 12m long
3rd Jan 1960	Marlboroigh		UAP 10m sphere
16th Apr 1960	Napier		UAP vertical cylinder
Dec 1963	Coromandel		"Other" Apelike
12th Jan 1965	Kaipara Harbour		UAP submerged
Jan 1965	Waihoaka		UAP landed & departed
3rd Feb 1965	Canterbury		7m UAP
24th Mar 1965	Muriwai Beach		10m "Other" hairy body
13th Nov 1965	Rugged Islands, Southland	Solid	2 UAPs out and in
20th Jan 1968	Wellington		UAP Large silver disk
2nd Feb 1968	Auckland		UAP Domed disk Killed man
July 1968	Pacific Oan		"Other" 2m monster in net
10th Sep 1968	Hawkes Bay		UAP 0.75m sphere
29th Nov 1968	Manawatu-Wanganui		3 UAPs Lights in triangle

NEW ZEALAND (Continued)

Date	Place	Credibility	Details
18th Dec 1968	Wairakei		1.7m "Other" in helmet
22nd Feb 1969	Awanui		3 "Others" 1.7m
31st Aug 1969	Gisborne		UAP Disk
4th Sep 1969	Ngatea		Several UAP nests
4th Oct 1969	Katikati		2 UAP Orbs
8th Oct 1969	Dargaville		4 UAP nests, Cigar sighted
30th Oct 1969	Waipukurau		UAP circular metal
Oct 1969	Watford		UAP nest
Oct 1969	Rotorua		3 UAP nests
Oct 1969	Takapuna		2 UAP nests
7th May 1970	Richmond		UAP giant saucer, at school
9th Apr 1971	Canterbury		UAP
14th Apr 1971	Taupo		UAP disk with dome
1971	Lyttleton Fishing		"Other" hippo sized in nets
15th July 1972	Hamilton		UAP 15m Disk 3"Others"
1972	Coromandel Range		"Other" tall & Hairy
8th Jan 1975	Auckland		UAP domed Disk
12th July 1975	Dunedin		"Other" 1.75m
5th Feb 1977	Auckland		2 "Others" human shape & size
April 1977	Gisborne		UAP domed disk "Others"
April 1977	Nr Christchurch		Fishers net 10m dead monster
July 1977	Kaiapot		"Other" big cat
7th Dec 1977	Gisborne		UAP disk
2nd Jan 1978	Auckland		UAP 2 "Others" 2.2m
30th May 1978	Auckland		UAP small cigar
16th Jul 1978	Auckland		UAP rectangle & 2
16th Dec 1978	Marlborough		UAP inverted wheatsheaf
1980	Gisborne		4 "Others" humanlike
11th Jul 1981	Dunedin		UAP football
12th Jul 1981	Canterbury		UAP half-oval
2nd Aug 1982	Canterbury		"Other" silver 2m
1st Jun 1994	Christchurch		UAP 4 lights
10th Jun 1995	Northland		UAP diamond
7th Feb 1998	Kapiti Island		UAP 4 green lights cylinder
26th Nov 1999	Auckland		UAP cigar

AUSTRALIA, PAPUA & NEW GUINEA

Date	Place	Credibility	Details
1821	Lake Bathurst		"Others" large Amphibians
1822	Sydney Cove		"Other" tall hairy Yowie
1828	Mount Wingen		UAP cigar
1844	New South Wales		"Other" tall hairy Yowie
1860	South Australia		UAP 18m orb
1872	Lake Burrumbert		"Other" Bunyip
1885	Victoria		"Other" Tiger-like
1885	Blue Mountains		UAP Disk 2 "Others"
1896	Bass Strait		UAP airship
July 1909	Blue Mountains		UAP cigar
15th Aug 1909	Perth		UAP airship
1910	Karumba		UAP disk abduction
Nov 1912	Currockbilly Mountains		"Other" 1.7m red hair, fangs
21st Jul 1913	Bathurst		"Other" 1.0m dwarf
1918	Watagan Mountains		"Other" 4m lizards
Jan 1924	Melbourne		UAP 6 Disks
1927	New South Wales		"Other" 3x long nose UAP
1930	Western Australia		"Other" 0.5m pink flesh
18th Jun 1931	Port Augusta Channel	Solid	UAP on surface
1933	Western Australia		UAP egg
Feb 1934	Queensland		"Other" 1.7m long arms
1938	Queensland		"Other" 1.8m Fair, no neck
Jan 1942	Tasman Peninsular		UAP Disk with dome
Feb 1944	Bass Strait		UAP Disk with dome
Summer 1944	Tasman Peninsular		UAP 30m long
6th Feb 1947	Port Augusta		3 UAPs egg shaped
Feb 1948	Berridale		UAP 3 small green "Others"
1948	Eggletons Plain		UAP 70m cigar
Oct 1949	Townsville		UAP 35m cigar
May 1950	Kings Tableland		UAP 14m egg
3rd May 1952	Melbourne		UAP silver cigar
18th Oct 1952	South Australia		UAP cigar 50m
1953	New South Wales		Several saucer nests
8th Jan 1953	Queensland		UAP 1m light
23rd May 1953	Melbourne		UAP 30m cigar

AUSTRALIA, PAPUA & NEW GUINEA Continued)

Date	Place	Credibility	Details
23rd Aug 1953	Port Moresby, New Guinea		UAP Dart shape
1953	New South Wales		UAP 9m disk
5th Feb 1954	Western Australia		UAP Disk
13th Mar 1954	South Australia		UAPs many multicolour
7th Apr 1954	Brisbane		UAP cigar
9th Apr 1954	South Australia		UAP long silver cigar
18th Apr 1954	Western Australia		UAP Disk with portholes
1st May 1954	Melbourne		UAP giant football
5th May 1954	Melbourne		UAP 30' cylinder
30th May 1954	Melbourne		UAP Oval size of a tram
4th Jun 1954	Melbourne		UAP green sphere
9th Jun 1954	Victoria		UAP 10m cylinder
5th Jul 1954	Melbourne		UAP cigar size of a car
31st Aug 1954	Over New South Wales		2 UAPs radar & pilot
9th Sep 1954	Victoria		UAP Disk
1954	Bankstown airport		UAP disk with portholes
Spring 1955	Balook		3 "Others" 23 cm tall
2nd Aug 1955	Eucla		"Other" Frog back
Aug 1956	Mount Pleasant		3 "Others" Cylinder UAP
1957	Bayswater		"Other" 1.4m
13th Jan 1958	Sydney		UAP huge dome
1958	Townsville		4 "Others" 2m fair
Jan 1959	Gulf Saint Vinnt	Solid	2 UAPs emerged
13th Mar 1959	Purnong Landing		UAP domed
26th Jun 1959	Papua & New Guinea		UAP over Mission
17th Mar 1960	New South Wales		UAP large candlestick
4th Oct 1960	Tasmania		5 UAPs, one large
6th Aug 1961	Western Australia		12 UAPs in pairs
20th Oct 1961	Sidney		UAP 60m long
1962	Bass Strait	Solid	UAP on sonar
16th Feb 1963	Moe, New South Wales		UAP Thick Disk
Spring 1963	Plympton		UAP disk 2 Tall "Others"
23rd Jan 1964	Gulf of Carpentaria	Certain	Pinwheel
16th Sep 1964	Tasmania		UAP attacks car
Jan 1965	Brisbane		100m Wide UAP
24th May 1965	Nr Nebo, Queensland		UAP circular, landed
9th Jan 1966	Tully, Queensland		UAP. Flying saucer nests
6th Apr 1966	Westall, Melbourne		UAP near school
24th Aug 1967	Woodonga, Victoria		UAP chased M/C
Dec 1967	Queensland		3 small "Others" in silver
3rd April 1968	Narrabeen Lakes		1.2 "Other" Elephant legs
10th May 1968	Tasman Sea	Strong	UAP emerged
1968	Brisbane		UAP Black Triangle
14th Jan 1969	Childers		UAP 3 giant "Others"
1969	Cairns		"Other" 2 tall hairy & child
1969	Tasmania		UAP Oval Tall, thin "Other"
8th Dec 1970	Western Australia		UAP silver disk, portholes

AUSTRALIA, PAPUA & NEW GUINEA Continued)

Date	Place	Credibility	Details
3rd May 1971	New South Wales		"UAP" entered sea
May 1972	New South Wales		3 "Others" 1.0m Tall
23rd Jun 1972	New South Wales		UAP cigar
3rd Jul 1972	Frankston Victoria		Large UAP saur
30th Apr 1973	New South Wales		UAP egg
Nov 1973	New South Wales		UAP 5m egg
6th Mar 1975	Tasmania		UAP Hexagonal
22nd Mar 1975	Nr Nebo, Queensland		UAP disk landed
19th Oct 1975	New South Wales		5 "Others" in UAP cigar
12th May 1976	New South Wales		UAP Disk crashed in sea
17th Oct 1977	Western Australia		"Other" 1.6m slender
Dec 1977	New South Wales		"Other" 1.5m no neck
21st Oct 1978	Bass Strait		Large UAP hit aircraft
Oct 1978	Queensland		"Other" 1.0m black hairy
21st Feb 1979	Queensland		UAP cigar
20th Oct 1979	Victoria		UAP in field beside road
29th Sep 1980	Gippsland, Victoria		UAP drained water tank
2nd May 1982	Queensland		UAP Disk. Many Nordics
Jan 1983	Western Australia		"Other" 1.0m
1983	New South Wales		"Other" Avian 2.0m
Feb 1984	Victoria		"Other" 1.5m no neck
Jul 1984	New South Wales		UAP 100m Ovoid
26th Dec 1984	New South Wales		4UAPS 1 cigar, 3 Disk
1985	New South Wales		"Other" 1.0m bulky
Jan 1987	Tasmania		"Other" 1.8m no neck
20th Jan 1988	Nullabor, W Australia		UAP attacks car
Aug 1991	South Australia		UAP Cone 1.3m "Other"
14th Oct 1991	New South Wales		UAP emerged
22nd Dec 1995	Western Australia		Abduction 2300miles 10min
Oct 1996	South Australia		UAP disk 70m
Dec 1997	Western Australia		"Other" Giant Flying

MALAYSIA & INDONESIA

Date	Place	Credibility	Details
16th Jul 1864	Gulf of Thailand	Moderate	2 Pinwheels
Oct 1891	South China Sea	Weak	Searchlight rays below
14th Mar 1907	Strait of Malacca	Moderate	Pinwheel
10th Jun 1909	Strait of Malacca	Solid	Pinwheel
12th Aug 1910	South China Sea	Moderate	Pinwheel
24th Aug 1911	Gulf of Thailand	Solid	Light out/in water
Jun 1930	Gulf of Martaban	Moderate	Pinwheel
1950	Malaysia		3 UAPs disks
23rd Oct 1952	Strait of Malacca		3 UAPs orange spheres
Sep 1958	Borneo		"Other" 6m
1959	Penang		"Other" Tiny man
1960	Perak		"Other" tiny family
1962	Penang		4 "Others" 1.7m
30th May 1962	Gulf of Siam	Strong	Pinwheel
1964	Pahang		2 "Others" tiny
1965	Perak		3-4 "Others" tiny
Aug 1966	Johor		"Other" 8m
1966	Selangor		2 "Others" 0.3m
2nd Jul 1969	Johor		5 "Others" tiny
19th Aug 1970	Penang		5 "Others" 7.5 cms
20th Aug 1970	Penang		2 "Others" 7.5 cms
24th Aug 1970	Selangor		5 "Others" 7.5 cms
1973	Perak		UAP tiny Several "Other" tiny
9th Oct 1974	Pahang		3 tiny "Others"
1976	Pahang		"Other" tiny
1978	Pahang		Several "Others" tiny
Mid 1979	Pahang		3 tiny "Others"
18th Jun 1980	Perak		3 tiny "Others"
24th Jan 1982	Perak		UAP Saturn shaped
21st Aug 1982	Sarawak		"Others" large heads 0.8m
Aug 1984	Selangor		4 "Others" large heads 0.8m
11th Oct 1985	Terengganu		Several "Others" 0.1m
27th Jul 1986	Terengganu		UAP 3m sphere. 1m "Other"
13th May 1991	Terengganu		"Hundreds" of tiny "Others"
Sep 1995	Selangor		UAP football pitch size
Dec 1996	Selangor		3 "Others" large heads
May 1999	Sarawak		Several tiny greenish "Others"

JAPAN & THE DRAGON'S TRIANGLE

Date	Place	Credibility	Detail
1942-45	Japan		UAPs Foo fighters
April 1945	Okinawa		UAP 12m cigar
15th Oct 1948	Japan		UAP chased by US fighter
21st Nov 1949	Akita		UAP rectangular
27th Aug 1950	Hachinohe		UAP flat 20m long
19th May 1950	Honshu		UAP Triangular
Dec 1950	Yellow Sea	Strong	2 UAPs entered
15th Feb 1951	Sea of Japan		5 UAPs on radar
9th Mar 1951	Japan-Korean Airspa		UAP radar
4th Mar 1952	Ashiya AFB		UAO disk
22nd Apr 1952	Okinawa		5 UAPs lights
30th May 1952	Oshima Island		UAP
14th Jul 1952	Okinawa		UAP sphere
5th Aug 1952	Haneda AFB		UAP on radar, sighted
14th Mar 1953	Hiroshima		90-100 UAPs
29th Dec 1953	Japan		UAP radar pilot sighting
12th Aug 1954	27°N 128°E		100 ft UAP
24th Mar 1955	Ryukyu Islands		UAP Disk with windows
15th Jan 1956	Sea of Japan	Strong	UAP entered
7th Sep 1956	Chiba		UAP disk – angel hair
9th Nov 1956	Yokohama		Several UAPs, thousands
17th Dec 1956	Itazuke		UAP on radar too fast
17th Jan 1957	Yokohama		UAP photo
21st Feb 1957	Yokohama		UAP fleet
Mar 1957	Le Shima Island		UAPJet crashed into saucer
13th Mar 1957	Tokyo		UAP cigar
19th Apr 1957	Pacific, near Japan	Solid	2 UAPs entered
10th Jun 1957	Yokohama		UAP cigar
20th Aug 1957	Fujisawa		UAP cigar, photo
10th Nov 1957	Lake Imba-numa		UAP cigar
26th Jan 1958	Shimada		UAP lands "Others"
2nd Feb 1958	Hokkaido		UAP egg shaped
17th Feb 1958	Yokohama		UAP cigar photo
26th Jan 1959	Hokkaido		2 UAPs
8th Apr 1961	34°N 132°E		Cigar UAP
21st Mar 1965	Osaka		UAP approached aircraft
11th Jan 1966	Western Pacific	Certain	Large UAP entered
15th Feb 1966	Osaka		UAP saucer
Sep 1966	Western Pacific	Solid	Light underwater
5th Jul 1968	Philippines	Solid	Conical UAP entered
1970	Mount Hiba		"Other" 1.3m hairy
Oct 1972	Hiroshima		"Other" 1.5m hairy
9th Jun 1974	Japan		UAP disk hit jet

JAPAN & THE DRAGON'S TRIANGLE

Date	Place	Credibility	Detail
23rd Feb 1975	Kofu		UAP landed 1.3m with fangs
15th Mar 2009	Japan		Several UAPs
16th Sep 2014	Japan		Triangular UAP

COASTAL RUSSIA

Date	Place	Credibility	Details
9th Aug 1845	Bornholm, Baltic Sea		Very Large Flame
16th Sep 1897	Arkhangelsk		UAP shining disk
30th Jun 1908	Tunguska, Krasnoyarsk		Massive explosion
Jun 1908	Krasnoyarsk		"Other" 1.2m dwarf
1920	Arkhangelsk		"Other" Hairy giant 6 toes
1947	Baltic Sea		Several Spherical UAPs
1950s	Black Sea		UAP emerged
16th Jul 1951	Murmansk		UAP disk 20m
Sept 1956	Sea of Japan		Enormous UAP crashed
Sep 1957	Black Sea		Spherical UAP
Jun 1958	Aral Sea		Disk UAP
1960	Barents Sea	Weak	UAP emerged
1961	Krasnoyarsk sky		UAP disk Passengers disappeared
Aug 1965	Black Sea		UAP Disk
Sep 1965	Kola Peninsular		2 UAPs one shot down other
1966	Black Sea	Weak	UAPs above navy
1967	Bay of Viborg		UAP Disk
Nov 1967	Baltic, nr Latvia		Gigantic UAP hemisphere
Aug 1970	Black Sea, Sochi		12 UAPs rectangles
1972	Krasnoyarsk		UAP sphere
7th Oct 1977	Barents Sea		9 UAPs circle ship
1978	Laptev Sea		UAP domed
28th Aug 1978	Kola Peninsular		3 "Others" tall
1979	Zapadnaya Sub Base		Disk UAPs overfly base
1979	Zapadnaya Sub Base		Disk UAP flies over dived sub
May 1979	Black Sea		Gigantic UAP Disk, exit
12th Aug 1979	Black Sea		100m UAP Sphere
Oct 27th 1979	Khatanga		Huge UAP & 3 spheres
17th Dec 1979	Khatanga		Multiple Sphere UAPs
End 1982	Black Sea Balaklava		UAP entered
Summer 1983	Black Sea, Bathyscape		Tall thin "Other" big eyes
Nov 1983	Kola Bay		3 UAP spheres
1984	Baltic		3m "Others" down 400m
Oct 1984	Laptev Sea		Radar Sphere UAP sighted
23rd Oct 1985	Arctic Oan		4 UAPs
July 1986	Arkhangelsk		UAP disk
11th Aug 1987	Black Sea		Triangular UAP
Sep 1988	Murmansk		"Other" Tall hairy
1989	Kil'din I, Olenegorsk		3 UAPs
Jul 1989	Black Sea		Cigar UAP & small square
2nd Aug 1989	Primorskiy Krai		Sphere UAP
2nd Nov 1989	Arkhangelski		UAP disk 45m
6th Nov 1989	Krasnoyarsk		"Other" 2.2m no neck
1989	Murmansk		"Other"
21st Mar 1990	Krasnoyarsk		2 huge UAPs

COASTAL RUSSIA (Continued)

Date	Place	Credibility	Details
Apr 1990	Kradnoyarsk		Several UAPs
Spring 1990	Franz-Joseph Land		UAP
Summer 1990	Murmansk		UAP disk 4 "Others"
1991	Arkhangelski		UAP sphere with ring
Nov 1992	Arkhangelski		UAP disk
Winter 1993	Odessa, Ukraine		UAP on surface
Summer 1995	Murmansk		"Other" Tall winged
13th Aug 1995	Karsk Sea		Submerged UAP
1996	Kirenniy3ni Cape		UAPs Spheres & Cylinders
1997	Black Sea		Bell UAP "borrowed" torpedo
Summer 1997	Cape Aiya, Crimea		2 very tall "Others" swimming
23rd Aug 1999	White Sea		"Other" pillar of light grew 200m
1999	Kola Peninsular		"Other" small

HAWAII

Date	Place	Credibility	Details
12th Aug 1825	At sea		UAP Red sphere in and out
18th Jun 1944	Oahu		UAP sphere 7 "Others 1.5m
18th Oct 1948	Oahu		UAP ellipse 5m
17th Jan 1950	Kauai		UAP disk
Apr 1950	Kauai		UAP 6 red lights
14th Mar 1952	From aeroplane		2 UAP disks
6th Mar 1953	Naval Air Station		75 UAPs
7th Jan 1956	Honolulu		UAP disk 35m
12th Mar 1963	Oahu		UAP Red sphere
6th Jun 1970	Maui		Many UAPs
15th Jul 1972	Kauai Naval Base		UAP forced down by Tomcats
1972	Maui		"Others" Nordics & Helmeted
1973	Wahiawa		"Others" Bigfoot, Giant woman
1974	Maui		Woman abducted permanently
1977	Maui		UAP attempted abduction of 4
1977	Maui		UAP 350m saucer
Jul 1980	Oahu		"Other" Dwarf
Aug 1980	Oahu		UAP mother ship & smaller
4th Sep 1994	Oahu		UAP 10x size of Aircraft Carrier
1996	Oahu		"Other" Menehune 1.0m

NOVA SCOTIA, LABRADOR & NEWFOUNDLAND

Date	Place	Credibility	Details
8th April 1813	Nova Scotia	Solid	UAP on surface
15thSept 1850	Wellington Channel, Canada	Strong	UAP above surface
12th Nov 1887	Cape Ra, Newfoundland	Solid	UAP emerged
Aug 1914	Georgian Bay, Canada	Moderate	UAP emerged
1932	OveeGreenland		UAP hexagonal disk
9th Jul 1947	Grand Falls, Newfoundland		4 UAPs round
10th Jul 1947	Harmon Field NFL		UAP disk
14th Aug 1947	Harmon Field NFL		2 UAP crescents
28th Oct 1948	Goose Bay Labrador		2 small UAPs on radar
21st Apr 1950	Cape Cod		UAP overtook jet
30th Jul 1950	St Johns NFL		UAP 3m long
4th Aug 1950	S of Nova Scotia	Certain	UAP above surface
10th Feb 1951	Nr Newfoundland		UAP emerged
Early 1952	Goose Bay Labrador		UAP sphere
April 1952	Labrador Sea	Solid	Sounds underwater
18th Aug 1952	Gander		UAP near miss
29th Aug 1952	Thule Greenland		3 UAPs silvery disks
23rd Oct 1952	Seal I. Nova Scotia		UAP light in water
24th July 1953	Greenland		UAP hits balloon
12th Oct 1953	Cape Cod		UAP destroyed jet
10th Nov 1953	Greenland		Large UAP
28th July 1954	Cape Cod	Certain	UAP emerged
24th Sep 1960	North of Labrador		UAP Cylinder entered
6th Oct 1961	Newfoundland Nuclear Site		UAP 4 minutes
29th Nov 1965	Cooper Creek, Nova Scotia		UAP dome with portholes
21st Nov 1966	Prince Edwards Island		UAP shiny disk
Summer 1967	New Brunswick		"Other" Knee-high
15th Sep 1967	R Cornwallis, Nova Scotia		UAP disk 4.5m
4th Oct 1967	Shag Harbour, Nova Scotia	Certain	UAP crash landed
11th Mar 1968	Marshalltown, Nova Scotia		UAP Oval
4th May 1968	Clark's Harbour, Nova Scotia	Solid	2 UAPs repairing
15th Sep 1968	Halifax, Nova Scotia		UAP domed 4.5m
25th Nov 1970	Shag Harbour, Nova Scotia		UAP lights
Sep 1973	Oak Island, Nova Scotia		"Others" 1 small, 1 taller
28th Jun 1975	Halifax, Nova Scotia		UAP 7m disk
3rd Aug 1976	Halifax, Nova Scotia		UAP 10m Oval
26th Oct 1978	Labrador		UAP Circular 2 hours
21st Oct 1983	Godhavn, Greenland		UAP oval dome
Summer 1987	New Brunswick		"Other"0.25m swimming
25th Jan 2010	Harbour Mile		Multiple UAPs
4th Jun 2014	Prince Edward I		Lights 22 mins

BERMUDA, FLORIDA AND THE SOUTHERN EAST COAST OF USA

Date	Place	Credibility	Details
27th Aug 1885	Bermuda		UAP triangular
29th Aug 1890	Nr N Carolina	Solid	UAP above
Feb 1894	W Coast of Florida	Moderate	UAP emerged
9th Feb 1913	Hamilton		UAP train heading south
27th Jun 1944	St Mary's Maryland		"Other" large winged
18th Jul 1945	Miami Naval Air		Plane missing
5th Dec 1945	Miami Naval Air		Flight 19 missing
5th Dec 1945	Miami Naval Air		Flying Boat missing
Jul 1946	Sussex Delaware		UAP disk in and out of sea
18th Nov 1948	Prince George Maryland		UAP highly manoeuvrable
28th Dec 1948	Miami Airport		DC3 disappeared
1950	Duval		UAP 10m 3 "Others" grays
4th Aug 1950	39°N 72°W		UAP circular
11th Jun 1952	33°N 75°W		UAP small square
16th Jul 1952	Wicomico, Maryland		UAP Bright light
Sept 1954	Nr Georgia	Solid	UAP emerged
22 Aug 1957	Miami Naval Air		UAP 15m bell
July 1959	North Carolina		Loud explosion in sky
24th Aug 1959	28°N 85°W		2 white lights
30th Sep 1959	36°N 70°W		Stationary light
1959	USS FDR Florida	Certain	UAP oblong
6th Aug 1960	29°N 79°W		White/green light
11th Jul 1961	30°N 76°W		Diamond UAP
15th Jan 1962	38°N 74°W		9 UAPs
29th Jan 1962	32°N 67°W		5 UAPs
12th Aug 1963	Nr Miami	Solid	UAP emerged
23rd Jun 1964	Palm Beach		UAP 2m
Summer 1965	Nr Miami		UAP disk
Oct 1965	Worcester, Maryland		4/5 "Others" egg heads
6th May 1966	Dade County		UAP 30m
5th Jun 1966	Brevard County		2 "Others" UAP submerged
1st Aug 1966	Prince George Maryland		UAP fast light, on radar too
Aug 1966	Sussex, Delaware		UAP landed
9th Feb 1967	Newcastle, Delaware		UAP Disk with cupola
19th Feb 1967	Broward County		UAP oval
3rd Mar 1967	Broward County		2 UAPs cones
Mar 1967	USS Annapolis		UAP Cigar
20th Jul 1967	Brevard County		UAP domed disk 20m
20th Mar 1968	Nr S Carolina	Solid	UAP emerged
2nd May 1972	Wicomico, Maryland		UAP disk. "Others" 1.2m
8th Sep 1973	Chatham, Georgia		UAP light
18th Oct 1973	Savannah Georgia		"Other" small
Jul 1974	West Palm Beach		UAP disk
22nd Aug 1978	Southampton		UAP 10m disk
21st Feb 1979	Nr Miami		UAP top shaped

BERMUDA, FLORIDA AND THE SOUTHERN EAST COAST OF USA (Continued)

Date	Place	Credibility	Details
28th Sep 1980	Over Palm Beach		2 UAPs disk buzzed plane
30th Jan 1984	Oil Rig near Miami		UAP 70m domed disk
Aug 1988	Miami		"Other" tall
May 1993	Dade County		UAP 30m Green "Other"
16th Mar 1995	Palm Beach		Several UAP disks
Dec 1995	West Dade County		3"Others" tall blond
6th May 1999	Brevard County		UAP 100m oblong

GULF OF MEXICO & COASTLINE

Date	Place	Credibility	Details
14th Jul 1841	At sea		"Other" Long neck monster
13th May 1872	At sea		"Other" Long neck monster
21st Jun 1908	At sea		"Other" Long neck monster
6th Sep 1946	New Orleans LA		UAP Red ball
6th Jul 1947	Galveston TX		7 UAPs disks
27th Jan 1949	Manatee, Florida		UAP 40m cigar
1st Apr 1952	320km s off Louisiana		UAP fell into sea
6th Sep 1955	Levy, Florida		6 UAPs
1957	Deep under oil rig		"Others" green frog-like
13th Aug 1959	Brazoria TX		UAP very bright
1961	New Orleans LA		"Others" 1.2m
July 1962	Neus TX		UAP submerged
3rd Sep 1965	Brazoria TX		UAP 60m cigar
Sep 1965	Veracruz, Mexico		UAP 7m
6th Feb 1966	Jefferson TX		UAP 4m Tadpole
26th Jul 1966	Yucatan, Mexico		UAP large disk
6th Apr 1967	Okaloosa, Florida		UAPs over school
May 1967	Hernando, Florida		"Other" Tall hairy
June 1967	Veracruz, Mexico		"Other" Tall hairy
27th Mar 1968	At sea		UAP fiery ball manoeuvred
Aug 1968	Pinellas, Florida		UAP 2m diameter globe
21st Aug 1968	Hernando, Florida		UAP Large globe
28th Oct 1969	Mobile Alabama		UAP 15m disk
Jun 1970	Veracruz, Mexico		"Others" 1.0m abducted
March 1973	Veracruz, Mexico		"Others" 1.0m abducted
11th Oct 1973	Pascagoula MS		UAP egg 3"Others" 1.5m Abdu
16th Oct 1973	Gulport MS		UAP
6th Nov 1973	Pascagoula MS		3m Submerged UAP
Jan 1974	Harrison MS		UAP Large Ovoid
15th Dec 1974	Jefferson TX		UAP disk
11th Nov 1975	New Orleans LA		Large disk UAP. Tall "Others"
30th Oct 1978	Aransas TX		UAP Triangle
1978	Veracruz, Mexico		"Other" Large Bat
Autumn 1979	Veracruz, Mexico		"Other" 3m tall thin
Apr 1986	Cameron TX		UAP disk
Spring 1988	Pensacola, Florida		UAP crash dead "Others" grays
6th Jan 1989	Pensacola, Florida		UAP Orange ball
9th Apr 1989	Pinellas, Florida		7 UAPs disks 2-3m
Apr 1989	Veracruz, Mexico		UAP gray sphere crash
11th Aug 1989	Brazoria TX		"Others" lizards Abduction
Apr 1990	Santa Rosa, Florida		6 "Others" Abduction

GULF OF MEXICO & COASTLINE (Continued)

Date	Place	Credibility	Details
19th Jan 1995	Quintana Roo, Mex		3 UAPs fiery balls manoeuvred
27th Oct 1995	Veracruz, Mexico		2 UAPs collision
Nov 1995	Veracruz, Mexico		UAP disk with dome
Feb 1996	Tamaulipas, Mexico		"Other" short red hair claws
3rd May 1996	Sinaloa, Mexico		"Other" Giant bat
20th Sep 1996	Quintana Roo		3 UAPs orange lights
Dec 1996	Yucatan, Mexico		"Other" very tall and thin
August 1997	Hernando, Florida		2 "Others"
17th Nov 1997	Pinellas Florida Sea		2 UAP amber lights
13th May 1998	Okaloosa, Florida		UAP Triangle

PUERTO RICO, BERMUDA TRIANGLE & SOUTH CARIBBEAN

Date	Place	Credibility	Details
Jul 1942	Toa Alta		UAP 5m disk
Early 1946	Barrio Utuado		"Other" 1m tall
14th May 1952	Mayaguez		2 UAP orange spheres
6th Oct 1952	Cabo Rojo, Puerto Rico	Solid	UAP entered
31st Dec 1952	Ramsey AFB		UAP Orange sphere
1956-1960	Laguna Cartagena		Multiple UAPs
18th Jan 1956	Nr Venezuela	Solid	Large UAP landed
20th Feb 1956	Vieques Island		UAP 10m Disk
13th Mar 1963	Nr Puerto Rico	Moderate	UAP on sonar
11th Apr 1963	Puerto Rico Trench	Solid	UAP under
Summer 1963	Between Puerto Rico & Florida	Sold	UAP on sonar
Spring 1967	El Yunque		"Other" Tall thin
1968	Bermuda triangle	Solid	UAP on surface
June 1968	Nr Puerto Rico	Moderate	2 UAPs in|out sea
July 1968	El Yunque		"Other" Tall thin
31st Dec 1968	Vega Baja		"Other" 1.7m
Summer 1971	El Yunque		"Other" Tall
28th Oct 1972	Rio Pedras		"UAP" Large oval
Nov 1972	Culebra Island		"Other" 1.3m reptilian
20th Oct 1973	El Yunque		3 "Other" Canine
2nd Nov 1973	Pina		UAP Oval
3rd Jan 1974	Bayamon		UAP Disk
Dec 1974	El Yunque		1m "Other" trapped
Mar 1975	Moca		"Other" Large Bird
12th Mar 1975	Moca		UAP big as a house
18th Apr 1975	Pon		"Other" Dwarf
1975	Caguas		UAP Egg
Jun 1977	Quebada Granda		"Other" big head eyes
31st Aug 1977	Mediana Alta		UAP windows, 4 small
17th Jul 1978	Camuy		"O" 2 Grays & 1 Tall
Apr 1979	Cabo Rojo		"Others" 3 Small Grays
Dec 1979	El Yunque		"Others" Several Grays
Summer 1979	San Juan		"Others" Large Bird
1982	El Yunque		2 "Others" 1.0 m
20th Nov 1983	Sierra del Yunque		"O" family of 3 tall hairy
Nov 1984	El Yunque		"Other" Nordic
Sunmer 1985	El Yunque		"Other" 2 giant birds
9th Jul 1986	El Yunque		7 "Others" Nordic
31st May 1987	Laguna Cartagena		UAP Huge Saucer
Late 1987	Laguna Cartagena		5 "Others" Small Grays
28th Dec 1988	Laguna Cartagena		UAP Large Triangle
Jul 1989	El Yunque		Large hole in ground
24th Sep 1971	Laguna Cartagena		UAP Disk "O" 7 small G

PUERTO RICO, BERMUDA TRIANGLE & SOUTH CARIBBEAN (Continued)

Date	Place	Credibility	Details
Oct 1992	El Yunque		"Other" 1.1 silky hair
1992	Maracal		"Other" 1.1 silky hair
May 1993	Vieques Islanda		UAP Manta Ray 15m
1994	San Juan		"Other" Large bird
7th Apr 1995	Orocovis		"Other" Gray with crest
11th May 1995	San Juan		"Other" 1.0m bird
Aug 1995	Dorado		"Other" Chupacabra
Dec 1995	Caja de Muertos		Large UAP
Feb 1996	Dorado		"Other" Chupacabra
2nd Apr 1996	Adjunta		3 UAPS large disks
4th Aug 1967	Nr Venezuela	Strong	UAP disk emerged
27th Aug 1967	Nr Venezuela	Solid	3 UAP emerged
12th May 1998	US Virgin Islands		6 UAPs orbs
25th Jan 2006	St Maarten		2 saucer UAPs
23th Apr 2013	Rafael Hernandes Airport		UAP in/out
8th Jul 2014	Culebra Island		UAP
20th Dec 2014	Cayeu		2 UAPs
27th Apr 2016	Northwest Coast		2 UAPs
1st Oct 2016	Puerto Rico		Saucer UAP

CUBA & WEST CARRIBEAN

Date	Place	Credibility	Details
Feb 1947	Havana		UAP Airship
1947	Kingston Jamaica		2 UAPs silver disks
9th Mar 1950	Santiago		UAP Disk
7th Apr 1952	18°N 65°W		Large circular UAP
Summer 1957	Bacuranao	Moderate	UAP emerged
17th Sep 1958	16°N 76°W		Large UAP
Autumn 1958	Guantanamo Bay		UAP Large disk
9th Oct 1958	Nr Cuba	Strong	Large UAP Cylinder
16th Sep 1960	19°N 93°W		Disk UAP
Apr 1961	Bermuda Triangle		2UAPs 1 large, 1 small disk
1961	Remedios		3UAPs 1 large 2 small spheres
20th Nov 1964	18°N 66°W		Triangular UAP
1968-1969	Guantanamo Bay	Solid	Multiple UAPs emerged
Late Oct 1969	North of Cuba	Moderate	Massive UAP
17th Aug 1981	Las Villas		UAP cigar crashed
1984	Pinar del Rio		"Other" tall & Hairy
21st Dec 1993	Las Villas		Several "Others" winged men
1994	Guantanamo Bay		UAP Cigar
13th Dec 1995	Mayabeque		UAP spikey sphere

BRAZIL

Date	Place	Credibility	Details
26th Nov 1846	At Sea		8-10m UAP
9th Feb 1913	Cape San Roque		Skytrain of many UAPs
Mid Jul 1939	A spring in Minas Gerais		UAP 2 Tall "Others"
4th Dec 1949	Rio de Janeiro		Disk UAP 2 "Others" 1.65m abdu
28th Nov 1953	Rio Guapore – N Brazil	Solid	UAP 6 Human-like "Others"
9th Dec 1954	Linha Bela Vista		Disk UAP 2 "Others"
14th Dec 1954	Curitiba		3 UAPs
16th Jul 1956	Caraguatatuba		UAP disk out 2 tall "Others" fair
26th Jul 1956	Nr Rio de Janeiro, Brazil	Strong	2 UAPs
10th Sep 1956	Niteroi		Spherical UAP 2 "Others" 1.6m
15th Dec 1957	Floodplain by river		UAP DiskDome 7 "Others" 1.4m
Jan 1958	Lagoa Negra Lake		Disk UAP 2 Tall "Others" 3 1.4m
10th Jan 1958	Atlantic Ocean near Brazil	Moderate	Large UAP in/out
16th Jan 1958	Trinidade Island		5x 1 Disk UAP
1962	Canoas		"Other" heals crash victim 1.7m
31st Oct 1963	Peropava River, Brazil	Strong	UAP disk entered
26th Jul 1965	Carazinho		2 Ovoid UAP 5 1.5m "Others"
Oct 1965	Canhotiho		Tube UAP 2 "Others" 0.9m
26th Feb 1966	Agua Branca, Quipapa		Disk UAP 1 Tall, several 1.7m
16th Mar 1966	Nr Rio de Janeiro, Brazil	Strong	Oval UAP parachutes
1967	Dantana do Livramento		UAP carries VW Beetle 2km
Jun 1967	Sarandi		Sphere UAP 3 "Others" .8m Abdu
30th Jul 1967	Santa Catarina, Brazil	Certain	Cigar UAP dived
6th Feb 1969	Pirassununga		UAP disk "Other" 3x1.5m
12th Feb 1969	Pirassununga		2 "Others" 1.3m
27th Jun 1970	Rio de Janeiro,		UAP Disk landed on beach
30th Aug 1970	Rio de Janeiro, Funil Dam		UAP Guard shot at it, and was hurt
4th Apr 1971-74 C	Itaperuna		3 x Sphere UAP 3x1.0m "Others"
2nd Oct 1971	Rio de Janeiro,		UAP Disk
4th Jan 1975	Sao Luis		UAP Disk 11x1.3m Abductions
July 1977	Maranhao & Para		UAP waves lights & burns football
11th Mar 1978	Rio de Janeiro		UAP disk wave 12
16th Mar 1978	Pelotas		UAP Disk 3 tall fair "Others"
24th Nar 1978	Penalva		UAP Domed disk 3 "Others" 1m
12th May 1978	Pelotas		Dimed Disk UAP "Others" 1.1m
Jan 1979	Santa Cruz		Egg UAP
27th Jan 1979	Santa Cruz		UAP attempted abduction
Nov 1979	Santo Antonio		UAP attempted abduction
1981	Santa Catarina		UAP Reptile "Other" abduction
6th Mar 1982	Morenaho		Cylinder UAP over football game
Oct 1982	Antonina		UAP landed 3 "Others" 1.9m
12th Aug 1983	Maraponga lake		UAP Disk abduction, dreams
2nd Jul 1985	Curitibar		Abduction to Rio de Janeiro
Nov 1988	Carnaubinha		Sphere UAP 2 "Others" 1.5m

BRAZIL (Continued)

Date	Place	Credibility	Details
1991	Acari		UAP attempted abduction
May 1991	Campo Redondo		House sized UAP near abduction
5th Mar 1992	Ceara interior		Bus sized UAP 5 "Others" 1.3m
17th Apr 1996	Belem		2 UAPs 8 "Others" 2.0m+
19th May 1986	Rio de Janeiro		Multiple UAPs seen & Radar
21st Aug 1996	Brazilia dam		UAP
5th Oct 1996	Sao Jose do Norte		Giant pyramid UAP
15th Jun 1997	Lago Oeste		Daytime UAP 30cm diameter
20th Sep 2010	Ceara		Circular UAP, many lights
25th Oct 2010	Belem		UAP
14th Nov 2010	Ceara		UAP

ARGENTINA

Date	Place	Credibility	Details
1870	Buenos Aires		Lizard Man found in cavern
1897	White Lake Patagonia		Long-necked lake monster
1905	River inSantiago		Blond forecast floods
24th Dec 1930	Bahia Blanka		Light over several entities
1943	Lincoln, Buenos Aires		2 "Others" 1.2 m
1953	Pehuajo		UAP disk 1.5m "Other"
20th Sep 1954	Brasen		UAP disk 1,4 m "Other"
4th Dec 1954	Coroonel Pringles		UAP disk, Dwarf
Aug 1956	Balneario Norte		UAP disk Tall "Other" Abduc
1957	Las Juanitas		"Others" enter UAP Disk
Feb 1959	Bahia Blanca		Abductee taken 1,155 km
25th Dec 1959	Ezeiza		80 cm "Other"
1960	Villa Crespo		Lizard Men "Others" tunnel
1961	Mar del Plata		15 "Others" Abducting
Sep 1961	Rio Salado		UAP disk 2 1.5m "Others"
1961	Mar del Plata		UAP Disk 3 "Others"
May 1962	Pehuajo		UAP 3 short "Others"
1962	General La Madrid		UAP bald "Other"
1962	Mar del Plata		2 "Others" at baby's bedside
Feb 1963	Ranelach		"Other" with fangs attacking
Nov 1964	Lanus		Short "Other" big head no ears
20th Jul 1965	Quinas		UAP egg. 2 Nordics
20th Aug 1965	Mar del Plata		UAP 2 orange "Others"
22nd Oct 1965	Rancaua		2 short bighead "Others"
1967	Olavarria		2 Tall "Others"
Mar 1968	Altura Maipu		Abduction 6400 km to Mexico
Apr 1968	Mar del Plata		1 Tall "Other"
22nd May 1968	La Florida		"Other" 1m green, big ears
4th Jun 1968	Buenos Aires		1 Nordic "Other"
14th Jun 1968	Mar del Plata		UAP Triangle 2m blond "Other"
20th Jun 1968	Avellaneda		14 giant "Others"
3rd Jul 1968	Ollavaria		1 2m blond & 1 short "Other"
9th Jul 1968	La Plata		1 Tall "Other"
26th Jul 1968	Mar del Plata		3 "Others" Lights on helmets
26th Jul 1968	Ollavaria		UAP oval. 2 tall "Others"
30th Jul 1968	Mar del Plata		2 "Others" Lights on helmets
6th Aug 1968	Canuelas		UAP disk 1 Tall "Other"
31st Aug 1968	Bahia Blanca		1 Tall "Other"
8th Jan 1972	Pergamino		2 short "Others" big heads
16th Mar 1972	Azul		2 "Others" 1.2m & 1.7m
27th Aug 1972	Bahia Blanca		UAP "Other" with big head
4th Oct 1972	Burzaco		5-10 Tall "Others"
28th Nov 1972	Tres Arroyes		1 Tall "Other"
30th Dec 1972	Tres Arroyes		UAP 1 medium "Other"

ARGENTINA (Continued)

Date	Place	Credibility	Details
Jan 1973	Buenos Aires		1 Tall "Other"
Jun 1973	Villa Martelli		1.7m reptilian "Other"
Summer 1973	El Cholo		UAP 3 0.7m "Others"
29th Oct 1973	General Pinto		UAP 3 Nordic "Others"
Fall 1973	Carmen De Arco		2x1.5m "Others"
15th Jan 1974	Mar del Plata		5 "Others" in masks exit sea
25th Jun 1974	Tandil		Tall "Other" with small head
9th Jul 1974	Tres Arroyas		Tall "Other" Lizard
5th Jan 1975	Bahia Blanca		Tall "Other" small headed
23rd Nov 1977	Capital Federal		2 "Others" 1.4m green
Jan 1978	Buenos Aires		Tall "Other"
28th Jun 1978	Sarandi		"Other" 1.4m pointed ears
31st Aug 1978	Necochea		2 "Others" 0.7m
Oct 1978	Road in Buenos Aires		UAP 4 Tall "Others"
28th Jan 1980	San Carlos de Bariloke		UAP Disk 3 Tall "Others"
Jan 1980	Mar del Plata		UAP 3 "Others" 1.7m lights
5th Jun 1980	La Plata		UAP cigar 7 "Others" lights
2nd Aug 1980	Mar del Plata		UAP floating 1.7m "Other"
1982	Lincoln		1.2m "Other"
1982	Veronica		UAP 1.8m "Other"
Summer 1982	San Miguel del Monte		2.5m Tall "Other"
25th Dec 1982	La Plata		2 small "Others"
1983	Capital Federal		3 small "Others" Big Eyes,
Jan 1984	Ezeiza		2 tall "Others" flying
Jan 1984	BA Proven		Vaporous "Other" Tall green
3rd Apr 1984	Barrio Coghla		10 "Others" 0.6m UAP sphere
12th Dec 1984	Pilar		4 "Others" 0.5m flying
Feb 1985	Azul		2 "Others" 0.5m
7th Dec 1985	Buenos Aires		Tall "Other"
1985	Buenos Aires		"Others" 6' felines x 100
Feb 1986	Don Torcinato		"Other" Giant bat with horns
Summer 1986	Benavidez		Tall Blond "Other"
Jan 1988	Mar del Plata		UAP Oval "Other" 0.5m
4th Feb 1988	Necochea		2 "Others" Tall blond
24th Aug 1988	Mar del Plata		2 "Others" 0.6m
14th Oct 1988	Pergamino		5-7 "Others" 0.7m
6th Nov 1988	Pergamino		6-7 "Others" Gray Big head eyes
1988	Mar del Plata		"Other" 1m
Feb 1990	Punta Indio		2 "Others" 0.9m gnomes
Mar 1990	Villa Devota		"Other" 0.8m big head, eyes
31st Aug 1990	Berisso		2 "Others" 1.0m
Sep 1990	La Plata		2 "Others" 1.0m
Feb 1991	Lake in Patagonia		"Other" Tall
Nov 1991	Buenos Aires		"Other" Tall blond
Jun 1992	Villa Devota		"Other" 0.8m large head

ARGENTINA (Continued)

Date	Place	Credibility	Details
Feb/ May 1992	Villa Devota		"Other" big head, eyes
Dec 1992	Colonia Urquiza		2 small "Others"
Jan 1994	Lanus		Small "Other"
24th Apr 1994	Olavarria		19 "Others" Tall
1st Jan 1994	Bandero		"Other" Yeti-like
11 Feb 1995	San Antonio Ouest		2 "Others" 1 short, 1 heavy-set
15th Nov 1995	Mar Del Plata		3 "Others" Large
Feb 1996	Carlos Spigazzini		"Other" Canine biped with claws
May 1996	Rio Parana		"Other" Lizard man
19th Mar 1999	Moreno		"Other" Short big head & eyes
Jul 1999	Merdes		"Other" Orange skinned man
25th Aug 1999	Trenque Laquen		5 "Others" Tall 2.5m
Nov 1999	Trenque Laquen		"Other" Tall

SOUTH AFRICA

Date	Place	Credibility	Details
18th Jan 1914	Transvaal		UAP Cigar searchlight
19th Aug 1914	Orange Free State		UAP Cigar searchlight
21st Aug 1914	Mafeking		UAP Cigar searchlight
Sept 1914	Natal		UAP
Oct 1928	Transvaal		2 UAPs spheres
Summer 1946	Johannesburg		UAP Disk 2 "Others" 2.1m
21st Feb 1950	Durban		UAP cigar
27th Dec 1954	Natal		UAP Disk windows 20m
25th Jun 1960	Eastern Cape		UAP 9m long
14th Dec 1963	Transvaal		UAP 15m disk
15th Sep 1965	Transvaal		UAP Top 10m
2nd Jul 1972	Eastern Cape		UAP triangle very bright
5th Jul 1972	Eastern Cape		UAP red disk
16th Jul 1972	Orange Free State		UAP 1m disk
22nd Jul 1972	Natal		UAP 10m sphere
19th Aug 1972	Natal		"Other" tall hovering
20th Aug 1972	Eastern Cape		UAP Big silver sphere
21st Aug 1972	Natal		"Other" 3m
28th Aug 1972	Eastern Cape		UAP multiple lights
12th Nov 1972	Eastern Cape		UAP sphere
17th Nov 1972	Eastern Cape		2 "Others" 1.0m tall red
19th Nov 1972	Eastern Cape		2 UAPs domed
16th Jan 1973	Eastern Cape		UAP 30m
28th Jun 1973	Transvaal		UAP domed
31st Jul 1975	Eastern Cape		UAP like caravan
5th Mar 1976	Eastern Cape		UAP bright disk
2nd Oct 1978	Cape Province		Oval UAP 20m "Other" gray
24th Nov 1978	Cape Town		UAP large egg
3rd Jan 1979	Gauteng		UAP 6m egg "Other" 1.5m dark hair
Summer 1986	Durban		20 "Others" 1m
Apr 1993	Western Cape		UAP oval
30th Mar 1995	Transvaal		UAP disk hole in it
14th Jan 1997	Orange Free State		UAP gray sphere
6th Jun 1998	Durban		UAP oval
1998	Durban		Tall bald "Other"

PORTUGAL

Date	Place	Credibility	Details
29th May 1946	Lisbon		UAP sighted
24th Sep 1954	Castelo Branco		2 "Others" 2.2m giants
15th Oct 1954	Alentjo		2 "Others"
28th Aug 1957	Centro		UAP disk with dome
2nd Nov 1959	Alentejo		2 UAPs angel hair
22nd Apr 1960	Alentejo		UAP cigar
1962	Algarve		Tunnels found
1st Jul 1965	Azores airport		UAP stopped clocks
Aug 1967	Minho		UAP ellipse
Sep 1973	Douro		UAP lens 2m 2"Others" 1.7m antennas on heads
1974	Algarve		UAP disk "Other" very tall
3rd Jan 1977	Beira Alta		"Other" like tall box on legs
14th Jan 1977	Amarante		"Other" egg shaped head
Sep 1977	Algarve		4 UAPs
6th Jan 1978	Estremadura		"Other" 2.3m
Aug 1978	Algarve		Several "Others" Tall
Sep 1978	Dao-Lafoes		UAP light
15th Oct 1978	Algarve		UAP round 2 tall "Others"
25th Dec 1980	Oeiras		UAP oval
2nd Nov 1982	Estremadura		UAP disk
25th Sep 1983	Douro Litoral		3 UAPs
1986	Almada		"Other" 2.0m
4th Sep 1989	Lisbon		Abduction

AZORES

Date	Place	Credibility	Details
10th July 1948	Azores		UAP sighted
26th Sep 1952	400 miles NNW of Azores		UAP 2 lights near plane
20th Sep 1954	Azores airport		UAP 3m, "Other" 1.8m
16th Aug 1956	Azores airport		UAP light near plane
Aug 1967	Azores		2 "Others" 2.2m
2nd Feb 1968	Azores		UAP Oval 4"Others"
15th Oct 1976	Azores		UAP circular

CANARY ISLES

Date	Place	Credibility	Details
1912	Tenerife		Passageway found
Sep 1969	Gran Canarias		UAP huge clear sphere
Summer 1973	Gran Canarias		UAP Box shaped
Jan 1975	Las Palmas		5 UAPs lights
Aug 1975	Gran Canarias		"Other" 1.4m big head & ears
21st Dec 1975	Tenerife		UAP circular
22nd Jun 1976	Gran Canarias		UAP Sphere 15m
Feb 1977	Tenerife		UAP 100m sphere emerged 4"Others" tall blond
11th Apr 1978	Gran Canarias		UAP cigar
Sep 1978	La Graciosa		"Other" very tall
3rd Mar 1979	Tenerife		UAP pear shaped
Spring 1980	Tenerife		"Others" tall thin
22nd Apr 1989	Tenerife		UAP Huge sphere entered sea
7th Feb 1990	Lanzarote		UAP huge red sphere landed
Dec 1990	Tenerife		UAP Disk
12th Oct 1992	Tenerife		UAP crash
Nov 1992	Tenerife		"Other" 1m furry cat eyes
1st Jul 1997	Tenerife		"Others" winged
7th Jul 1997	Tenerife		UAP 30m saucer "Others" 1.2m
Jan 1999	Tenerife		UAP domed disk, Tall, long-haired "Others"

NORWAY

Date	Place	Credibility	Details
Summer 1869	Sor-Trondelag		UAP sighted
13th Aug 1897	Melo		2 UAPs
1st Apr 1908	Notodden		UAP crashed Egg shaped
Summer 1915	Sulitjelma		UAP landed, bell. 2 "Others"
Jan 1934	Multiple sites		UAPs – Ghost Fliers
31st Mar 1934	Sandnessjoen		Huge UAP
28th Jan 1937	Oftofjorden		Submersible UAP
14th Mar 1942	Banak		UAP cigar, radar & visual
1946	Scandinavian Lakes	Certain	UAPs diving into lakes
Aug 1947	Lake Djupsjoen	Solid	UAP blue egg entered
1950	Malselv		UAP disk "Other" contact
20th Sep 1952	Operation Mainbrace		UAP sphere photo
Oct 1952	Lagen River	Moderate	UAP entered, saucer
28th Nov 1953	Nr Oslo		UAP disk follows car
13th May 1954	Norbotten		UAP sphere
20th Aug 1954	Mosjoen		UAP disk, Tall "Other"
Jul 1957	Halingdal		UAP in flight
4th Jan 1958	Stavanger		UAP landed, Tall "Other"
1st Jun 1958	Alta Fjord	Strong	UAP entered. Delta wings
23rd Dec 1959	Skomsvol		UAP cigar dropped 2 UAPs
24th Sep 1961	Baltic, Nr Poland	Solid	UAP emerged
Winter 1962	Lake Stordalsvatnet		UAP winged
28th Feb 1963	Atlantic, Nr Norway	Solid	UAP buzzed British ships
21st Aug 1963	Skjervoy		UAP oval. Taking in water
21st Aug 1965	Kvaenangen Fjord	Strong	UAP fishing wartime wreck
14th Dec 1966	Porsan Fjord		UAP entered
30th Mar 1969	Oslo		UAP follows car
6th Jun 1969	Husnes		UAP disk
Jan 1970	Tronstad		UAP disk
3rd Nov 1970	Southern Norway		UAP light blinds car driver
1st Jan 1972	Bergen Airport		7 UAPs seen by pilot
1st Jan 1972	Nedre Lerfoss Power Stn		UAP cigar shape
Summer 1972	Solvang		UAP disk
20th Nov 1972	Fjordane multiple		Submersible UAP
10th Aug 1974	Hamar		Submersible UAP
Nov 1975	Arendal		UAP dropped off 8 larva
7th May 1977	Undersaker		2 UAPs triangles
Aug 1977	Narvik		UAP disk, 2 "Others"
1981-2-3-4	Hessdalen		Multiple UAPs
Nov 1984	Skogveien		UAP contactee Arve
28th Oct 1985	Hengsle, Honefoss		Multiple small "Others"
9th Mar 1992	Hamar		UAP Small triangle
Oct 1992	Hardangervidda		UAP 2 short "Others"
12th Apr 1977	Hordaland		UAP small triangle
Oct 1998	Trondheim		Large UAP Tall "Other"

SWEDEN

Date	Place	Credibility	Details
Jun 1935	Dalarna		"Other" Little man
1941	Stockholm - King		3 "Others" 1.2m dwarfs
Spring 1943	Ostergotland		UAP disk
Aug 1943	Kristianstad		"Others" line of little ones
25th May 1946	Stockholm		Start Ghost Fliers barrage
11th Jul 1946	Vastagotland		UAP Sphere crashed
11th Jul 1946	Orebro		UAP 20m triangle
Aug 1946	Tarno		"Other"
28 Sep 1952	Dalarna		2 UAPs Cigars with windows
18th Aug 1949	Stockholm		UAP disk 15m
13th May 1954	Norrbotten		3 UAPs disk
8th Sep 1957	Vastagotland		2 "Others" Cyclops conical
9th Nov 1958	Uppland		UAP flattened sphere 17m
20th Dec 1958	Skania		UAP 5m disk "Others" slugs
29th Sep 1959	Smaland		UAP 4m oval "Others" teens
29th Oct 1965	Stockholm		5 "Others" small big heads
5th Mar 1967	Vastagotland		UAP cigar 25-40m long
23rd Aug 1967	Ostergotland		"Other" 1.3m small mouth
1967	Vojman River		UAP30m disk
15th Feb 1971	Norrbotten		"Other" 1.0m cap
30th Apr 1971	Jonkoping		UAP Disk 10m
3rd May 1973	Ostergotland		UAP Sphere
21st May 1973	Kristianstad		"Other" large black bird
22nd Jan 1974	Stockholm		Several "Others" 2m thin
14th Jan 1975	Ostergotland		UAP short wings
Aug 1975	Scockholm		2 "Others" short gray
7th Jan 1976	Jamtland		UAP pyramid
3rd Mar 1976	Nottbotten		7 UAPs disk
Aug 1976	Dalarna		"Other" dwarf v large ears
14th Nov 1976	Kolsva		UAP tropical helmet 6m
31st Jul 1980	Lapland		UAP cigar entered lake
1st Nov 1980	Orebro		UAP oval 12m
Summer 1981	Smaland		"Other" 1.0m on film
20th Aug 1983	Dalarna		UAP globe
Summer 1985	Helsingland		"Other" Tall & hairy
Oct 1993	Vaserbotten		UAP disk
17th Mar 1997	Vastra Gotland		UAP ball
27th Jul 1999	Varnland		UAP 5m oblong entered lake

RUSSIAN LAKES & RIVERS

Date	Place	Credibility	Details
1663	Robozero Lake, Vologda		Large flaming sphere UAP
19th Century	Lake Lagoda		Regular Underground Noises
1884	Lake Baikal		Large UAP Sphere
1930s	Lake Issyk Kul		3 Giant "Other" Skeletons
1938	Lake West Yaroslavl		Several "Others" large heads
1939	Lake Shaitan Kirov		"Others" Green 1,2m
1941	Zelyeny Island River Don		UAP dogfight 1 crashed 20m
Summer 1943	Lake Shaitan Kirov		"Others" Tall
16th Jun 1948	Lake Baskunchak		UAP cigar damaged jet
1950s - 1960s	Chersky, Kolyma River		5-7 Disk UAPs
1950	Chersky, Kolyma River		Disk UAP 3 days. stayed 2 hrs
Sep 52	Chulym River Khakassia		"Others" Short large heads
1955	Enisey River Krasnoyarsk		"Others" little green men, playing with children
21st Apr 1961	Lake Onega		UAP Egg shaped crashed
27th Apr 1961	Lake Onega, Karelia	Strong	UAP in and out
15th Jun 1962	Kabarga River		UAP red sphere
1963	Lake Labynkir, Yakutsk		"Other" Lake monster
17th May 1964	Lake Baikal		UAP sphere approaches post.
Autumn 1965	Lake Baikal		Gigantic cigar UAP
17th May 1966	Nizhnyaya River		UAP 30m cone 3 Tall "Others"
June 1966	Lake Kolyvansky		UAP 8m droplet, 3 "Others"
Winter 1966	Lake Ladoga, Karelia		UAP disk 8 humans killed
2nd Aug 1972	Vyatka River, Kirov		"Other" 2m Tall pointed ears
1973	Lake Onega		Triangular UAP
Jul 1974	Esaulovka River Siberia		2 "Others" Tall Abduction
Jul 1975	Charvak Reservoir		UAP sphere exited
Aug 1975	Kuril Islands		Abduction 2 weeks
Feb 1997	Lake Lagoda		Triangular UAP
May 1977	Lake Kemsk, Vologda		UAP fiery sphere exploded
20th Sep 1977	Lake Onrga, Karelia		UAP Disk 100m
30th Dec 1980	Valaam I. Lake Ladoga		"Other" Tall
Summer 1982	Lake Baikal		Tall "Others" underwater
Aug 1988	Lake Lovozero Murmansk		"Other" Tall hairy
1982	Lake Sarez		UAPs enter and leave
13th Feb 1989	Lake Kopansky, Leningrad		UAP 30m disk windows
June 1989	Lake Zumerkiy, Vologda		2 Gray-green "Others"
Aug 1989	Lake Lagoda, Valaam I		Large Spherical UAP
19th Jan 1990	Lake Khnaka, Primorsky		UAP Disk landed on lake

THE UFO, E.T, ALIEN TRILOGY

RUSSIAN LAKES & RIVERS (Continued)

Date	Place	Credibility	Details
Feb 1990	Lake Issyk Kul		Large UAP sphere split into 4
21st Mar 1990	Amur River		UAPs Red Sphere, Cigar
30th Apr 1990	Lake Torbeevo		UAP 5m Sphere
1990s	Lake near Leningrad		Hang-glider - 3 Big "Others"
20th Jun 1990	Lake Issyk Kul		Triangular UAP above
17th Aug 1990	Lake Malyi Byadan		UAP 6m disk 2 Tall "Others"
13th Sep 1990	Lake Vselug Tver		UAP Sphere
1994-96	Lake Sukhodol'skoye		Spherical UAPs
21st Jul 1997	Caspian Sea		UAP entered sea
1990-2000	Lake Cheremenetskoye		Frequent UAPs
Spring 2009	Lake Baikal		2 orange UAP spheres

THE GREAT LAKES

Date	Place	Credibility	Details
27th Jan 1895	Lake Michigan	Weak	Large UAP emerged
Summer 1950	Flambeau Lake, Wisconsin	Weak	UAP emerged
26th Aug 1952	Seneca Lake, New York	Moderate	UAP entered
July 1955	Lake Ontario	Solid	UAP emerged
17th Sept 1955	Ticitus Reservoir, New York	Strong	UAP above surfa
20th July 1958	Small Lake, Michigan	Certain	UAP entered
14th Oct 1961	Lake Superior	Certain	Cigar UAP entered
7th Sep 1962	Montrose		Domed UAP
15th Sep 1962	Oradell Reservoir, New Jersey	Certain	UAP landed on water
August 1964	Upper Nemahbin L, Wisconsin	Solid	UAP in and out
Jan, Oct 1966	Wanaque Reservoir, New Jersey	Certain	Multiple UAPs above
June 1966	Straits of Mackinac, Michigan	Solid	2 UAPs emerged
20th June 1969	Southern Lake Michigan	Certain	UAP in and out
29th Jul 1989	Ludington		5 discs emerge
22nd May 2005	Michigan		Large irregular shape
1st Mar 2009	Oakville		Massive UAP
12th Sep 2009	Lake Huron		Entered Lake
15th Aug 2011	Bay City		Dull rectangle
17th Oct 2012	Coloma		6 Orbs emerge
14th Sep 2014	Lake Michigan		3 Orbs leave lake
24th Mar 2015	Toronto		V-shaped 5-7 lights
22nd Aug 2015	Hamilton		Multiple Lights
28th Nov 2016	Green Bay		Triangular UAP
9th Apr 2018	Benton Harbour		Cigar Shape
5th Sep 2018	Wasaga		Black Triangle

NORTHERN ITALY

Date	Place	Credibility	Details
1845	Tuscany		UAP huge disk
1864	Tuscany		UAP huge fiery globe
13th Dec 1884	Lombardy		UAP crash
Jun 1905	Emilia-Romagna		"Other" Flying Woman
Jul 1910	Tuscany		"Other" Giant flying bat
Dec 1919	Veneto		"Other" Luminous Spectre
1927	River Po, Veneto	Moderate	UAP in & out
Summer 1929	Emilia Romagna		UAP orange ball
Summer 1929	Tuscany		UAPs multi orange balls
13th Jun 1933	Lombardy		UAP crashed & studied
28th Nov 1942	Piedmont		UAP 65m long
13th Mar 1945	Tuscany		100 UAPs orange balls
1st Sep 1946	Marche		UAP into sea
14th Aug 1947	Friuli		3"Other" 1m green 8 fingers
3rd Apr 1948	San Martino, Lombardy		UAP & "Other"- 1.7m'
28th Oct 1049	Marche		20m UAP cigar
31st Mar 1959	Emilia Romagna		EAP Egg into sea
24th Apr 1950	Lombardy		UAP & 4 "Others" Helmet
1951	Piedmont		Huge UAP
31st Jul 1952	Musinè		UAP landed
18th Nov 1952	Venrto		UAP Disk with dome
19th Sep 1954	Rome		UAP Cigar
18th Oct 1954	Lombardy		UAP Short "Other"
19th Oct 1954	Tuscany		UAP 2 "Others"
27th Oct 1954	Tuscany Football Stad		Cigar UAP over thousands
28th Oct 1954	Rome		3 UAP flying fast
6th Nov 1954	Rome		Fleet UAPs
8th Nov 1954	Lombardy Stadium		UAP "Others" helmets
14th Nov 1954	Liguria		Cigar UAP 3 "Others"
15th Nov 1954	Rome		Fast UAPs
Sep 1955	Emilia Romagna		"Other" 1.0m
Apr 1956	Marche		2 "Others" 2.5m 1.0m
Jun 1958	Lombardy		UAP Box
15th Mar 1959	Turin Airport Piedmont		UAP Disk
3rd Jun 1961	Nr Savona, Adriatic Sea	Solid	UAP emerged
9th Apr 1962	Piedmont		UAP Disk
9th Nov 1962	Emilia Romagna		UAP landed 2 "Others"
17th Dec 1962	Milan		UAP landed 2 "Others"
4th Jan 1963	Rome		UAP disk with dome
20th Aug 1963	Rome		UAP domed disk
1st Nov 1964	Mount Musinè		UAP Landed
Jul 1967	Lombardy		UAP Disk 6 "Others"1.2M

NORTHERN ITALY (Continued)

Date	Place	Credibility	Details
15th Nov 1967	Tuscany		UAP delta shaped
22nd Aug 1968	Latina		UAP Disk with dome
24th Sep 1970	Piedmont		Triangular UAP
4th Jul 1973	Liguria		Round UAP ball of fire
30th Nov 1973	Turin Airport, Piedmont		Large UAP
5th Dec 1973	Piedmont		UAP radar and thousands
16th Apr 1974	Piedmont		UAP Ring with Cockpit
15th Feb 1975	Tuscany		2 "Others" Tall
Jun 1975	Emilia Romagna		"Other" Giant lizard
13th Jan 1976	Liguria		4 "Others" 1.6m
28th Mar 1976	Marche		Oval UAP left sea
14th May 1976	Liguria		"Other" 1.3m
10th Sep 1976	Liguria		"Other" 1.2m
1st Jul 1977	NATO Base Friuli		UAP bright light
6th Oct 1977	Tuscany		3 UAPs half moon size
23rd Apr 1978	Tuscany		"Other" 1.0m
24th Oct 1978	Marche		Large UAP up and down
18th Dec 1978	All Over Italy		Multiple UAPs
31st Dec 1978	Marche		UAP sphere up from sea
16th Jan 1979	Piedmont		10m UAP sphere
3rd Feb 1979	Liguria		"Other"1.2m helmet light
7th Feb 1979	Lombardy		"Other" 2.0m long arms
4th Oct 1979	Viggiu, Liguria		2 "Others" dwarfs <1m
14th Jun 1980	Liguria		2 "Others" 2.0m
8th Nov 1981	Veneto		UFO oval 2m "Other" no neck
Jul 1983	Tuscany		UFO 60m sphere
9th Oct 1984	Tuscany		Elliptical UAP. "Other"
7th Jun 1986	Piedmont		"Other" Large Red Hair
Summer 1986	Veneto		"Other" Tall Reptilian
15th Aug 1986	Veneto		UAP Domed disk
27th Sep 1986	Emilia Romagna		UAP Huge with wings
28th Oct 1986	Tuscany		UAP large triangle
16th Feb 1988	Piedmont		UAP disk "Other" 2.0m
20th Apr 1992	Veneto		UAP cigar 2"Others"
Jun 1993	Marche		"Other" Dwarf big feet
10th Jul 1993	Lombardy		5"Others" Luminous
3rd Feb 1994	Marche		"Other" 1.5m big feet
22 Jul 1996	Veneto		"Other" 1.0m
9th Nov 1996	Veneto		3 "Others" small gray big head
1st May 1998	Lombardy		"Other" big hairy
18th Sep 1998	Liguria		UAP blue oval
6th Jun 2010	Vatican City		Triangular UAP
27th Apr 2013	Province of Verona		UAP

PART 3

EARTH'S ALIEN SYLLABUS

THE UFO, E.T, ALIEN TRILOGY

Table of Contents

	PREFACE	209
1	**THE 50,000 YEAR PLAN**	211
	a) I Get Convinced UAPs Are Real	211
	b) The Anomaly	212
2	**40,000 YEARS LATER**	213
	a) The Grand Plan	213
	b) Agriculture	213
	c) Building Stone Structures	213
	d) Metallurgy, Mining & Blacksmithing	215
	e) The Greek Classical Period	217
	f) China	218
3	**THE LAST THOUSAND YEARS**	221
	a) The Renaissance	221
	b) The Railways	223
	c) Chemistry	224
	d) Electrical Power Generation & Transmission	226
	e) Telecommunications	228
	f) Chinese Development	229
4	**THE LAST CENTURY**	231
	a) The Development of US Society	231
	b) Power Transmission	232
	c) Extra-Terrestrial Technology	234
	d) The "Other" Team	235
	e) An Alternative Syllabus	236
	f) The Majestic Twelve	237
	g) Current United States' Activities	239
5	**THE WAY FORWARD**	243
	a) Spaceflight	243
	b) Trading with "Other" Colonies	244
	Appendices	247

PREFACE

In the first two parts of this book "UFO – Friend or Foe?[319]" and "We Own 29% - ET Has the Rest", [320]I have made a broad-ranging sweep through the history of UAP and "Other" sightings across as much of our planet as I could.

In reviewing evidence of all the sightings of UAPs, I was easily able to convince myself, and hopefully my readers, that they really existed, and that they had some bases underground, and underwater.

President Dwight Eisenhower made a grave mistake when he signed a deal with the Small Grays where, in return for information about "Other" technology, he allowed them a base in the US and granted them permission to take cattle and humans for experimental purposes, provided that the humans were returned unharmed, and they limited the numbers that they took. The Small Grays, once they were comfortably installed, have broken every part of the agreement, and the US can do nothing about it. In addition they have stopped the US (and Possibly Russia) from keeping their nuclear weapons at readiness, and have insisted on their presence being kept secret until they authorise disclosure.

There is evidence that a group of "Others" have been nurturing the human race for tens of thousands of years, and they did warn the US against signing a deal with the Small Grays.

In Part 2 of this book[321], I showed that there are a number of "Other" races who have established small colonies on Earth, together with many bases. Some of these may be the same "Others" who have been acting to encourage the development of the Human Race. It will be interesting to see whether they come into conflict with the US Military-Industrial entity known as Majestic-12.

I showed that Majestic-12 was developing spacecraft, but it is difficult to determine how advanced these are, or the uses to which they have already been put.

Chapter 1

THE 50,000 YEAR PLAN

(You Can Forget Those 5-year Plans)

a) <u>I GET CONVINCED UAPs ARE REAL</u>

Ancient clay tablets[322] found in Sumeria dating back thousands of years suggest that Humans were visited by strangers called the Anunnaki, a race of giants that they worshipped as Gods. When viewed with a modern eye, these would probably be considered as some form of Extra-Terrestrial.

In Part1 of this book, I reviewed a wide range of UAP sightings to try and get a feel for their likely veracity. I soon excluded any sightings where there was only one witness on the basis that I was not able to go into their possible motivations, whether honest, dishonest or delusional. Then, I eliminated mysterious cloud formations and single unmoving points of light, on the basis that they could so easily be miss-identifications.

What remained was a group of sightings which could not be explained easily. If it wasn't so worrying, it would be comical: there were desperate attempts by officialdom to account for these, including such fantasies as marsh gas burning[323] high in the air rather than getting diluted as it rose, or the highly improbable chance coincidence[324] of a meteor and an earthquake in a non-earthquake zone.

Then we have threats on witnesses and their families, either of violence or of destruction of their careers.

Even if the evidence of these remaining sightings were not so convincing, the desperate attempts at denial would probably have convinced me of the reality of UAPs and "Others".

I remain certain that UAPs and "Others" exist, even if they have been in our folklore as far back as human memory can stretch, and I hope that I managed to convince some of my readers in my first book. I accept that there are sincere skeptics who cannot believe, but I invite them to read further to see whether some strange events are better explained in my interpretation, than in the "Official" version. Otherwise, just read and enjoy it as a fantasy.

b) THE ANOMALY

Some Geneticists say that the human race has developed faster than other animals in the period since the extinction of the dinosaurs. For a long time, they all developed at the same rate then, suddenly, about 50,000 years ago, there was a sudden sprint in human brain development, which may have, in turn, triggered an explosion of creativity and invention.

In 2006, geneticist Dr Coleen Clements from Rochester University, published a book[325] showing that, at some time in history between 60,000 and 14,000 years ago, something called the "D allele" entered the human microcephalin gene of our DNA[326]. This had the effect of encouraging the growth and development of the human brain[327].

The usual way in which this might happen is by interbreeding, but tests[328] have shown that neither of the possible candidate hominids, Neanderthals or Denisovians, had this either, so it must have come from somewhere else. She proposed that it was introduced by "Others". This was controversial, but has not been disproved.

If any of our friendly skeptics are still with me, the challenge is for you to think up an alternative.

It is suggested that a race called the Anunnaki[329] were responsible for this modification to our genes. They had discovered an inquisitive ape on this planet, ranging across the whole world which, unlike Neanderthal, Flores, Denisovian, or other strains of hominid around at the time, they considered could be developed into a contributing member of their confederation. If true, this plan reshapes our whole understanding of humanity's purpose.

Chapter 2

40,000 YEARS LATER

a) <u>THE GRAND PLAN</u>

It seems amazing to us that the Anunnaki thought so long-term. About 10.000 years ago, the Human Race had developed to the level where they could commence the next stage of their work. Many tribes around the world have folklore describing how "Gods" came to teach them. This happened in North, Central and South America, Europe, India, Australia and New Zealand. The descriptions of the "Gods" vary, but they were all doing the same thing. Seen with modern eyes, these "Gods" were "Others".

b) <u>AGRICULTURE</u>

About 12,000 years ago, in Mesopotamia for example, it all started with a human tribe of hunter-gatherers which was eking out a living on the banks of the Euphrates. Clay tablets[330], found by modern archaeologists, describe how a species called the Anunnaki, giant humanoids, possibly assisted by Avians[331], a bird-like species, arrived there and started introducing them to ways of making their lives easier. The humans learnt agriculture, mathematics, astronomy, writing, schooling and the calendar. They were the first to introduce the 24 hour day and the 60 minute hour. It was also during this time that the wheel was invented.

All these innovations are important, but that of agriculture was crucial. By learning about irrigation, cattle husbandry, the gathering and sowing of seeds and rotation of crops, they could produce more food for the same effort. Alternatively, it provided spare time for other activities.

Pictures of Anunnaki and Avians appear in Egyptian paintings and carvings, and this suggests that they were spreading their tuition widely. They would probably have had dealings with the early Persians, Turks and Greeks as well.

It is recorded that the Anunnaki left Earth about 5000 years ago, by which time all these innovations were spreading around the world.

c) BUILDING STONE STRUCTURES

One of the more contentious points in archaeology is whether early civilizations such as the Mayans or Egyptians had the capability to cut and move these massive stones, and erect these colossal edifices. It would certainly have been easier for them if they had received the advice or assistance of "Others".

Tales of the construction of Stonehenge describe the stones being moved by magic (or music?). The islanders of Pohnpei in Micronesia say the same about Nan Madol[332]. In South America, it has been said that the ancients had a way of softening rock to shape it.

We'll never know, but this could have been the first step in creating our civilization. Certainly someone like the "Other" race known as Dwarfs, small people famed for their stone and metal working skills, (See Chapter3 section d) would have had the capability to work with edifices like these, and it is possible that they did exactly that.

The question is twofold –

- Who built these monolithic structures?
- Why?

Pyramids have turned up all over the world, but we don't know whether they were the result of parallel development, or simply copies, assuming humans from Central America were capable of visiting Egypt at that time. Examples are:

- Egypt – obviously   2650 BCE
- Baghdad[333]   1500 BCE
- China[334]   2300 BCE
- Central America[335]   900 BCE

It would strange if the Star Trek's "Hodgkin's Law of Parallel Planetary Development" was actually applicable!

I am left wondering whether the idea, and possibly the design, for these comes from "Others" rather than humans, even if the execution was purely human.

It is the spread of pyramid sites around the world which is my challenge to the skeptics here.

d) METALLURGY, MINING & BLACKSMITHING

Once humans had spare time from scraping together enough food to survive, ideas could be introduced to encourage their curiosity and creative urges.

Copper had been worked for a long time during the Stone Age but, about 3,000 BCE (5,000 years ago by coincidence), tin was first discovered in Anatolia in Turkey[336]. Very soon after that, tin was added to copper to make bronze, and the Bronze Age began[337], again in Anatolia. Who thought to do that? It certainly advanced civilization significantly.

It was not long before the demand for tin outstripped the supply around the Mediterranean[338], and skilled miners, most probably Dwarfs, started looking for new sites, heading westward to Spain and Portugal, northwards to Britain and Germany, and eastwards to India, passing on their knowledge and craft to the locals and thereby earning their reputation which lasts to this day. Appendix 4 contains information on Dwarfs. The Bronze Age started in Britain in about 2,000 BCE.

However, the Anatolians were not resting on their laurels. In 1,300 BCE, they were the first to produce iron[339], from ore as opposed to using meteoritic iron, initiating the Iron Age, and gradually bringing the Bronze Age to an end. This happened in Britain in about 800 BCE.

The Anatolians were clearly feeling unappreciated so they became the first to produce high carbon steel[340]. Even into the Common Era, the manufacture of Damascus steel required skilled forging technique to produce very high quality swords. This was invented in India, perhaps by Dwarfs working there.

Toledo sword manufacturing in Spain, during the same period, used the same forging principles as for Damascus steel, but the changes in weaponry proved the death knell for many of the artisan manufacturers[341]. However Spain still has a flourishing steel industry with a reputation for fine knives. There were indications of a Dwarf base in the north[342], near Barcelona or the Basque country, where the first Spanish iron and steel were produced, but they appear to have left in the 1960s.

In Pittsburg in the United States, there were a number of multiple sightings of Dwarfs between 1905 and 1956[343]. Since the 1960s, there have been many claims of Small Gray abductions, and many sightings of UAPs and of Bigfoot, but only one of a Dwarf in 1965. The city is close enough to Lake Erie for there to have been an "Other" base there, or they could have set up a now-abandoned base in a nearby lake.

In Northumberland in England, lived the Simonside Dwarfs[344] also known as Brownmen, Bogles, or Duergar. They appear to have moved away in the 1970s, judging from the sighting reports[345]. One of their parting shots in guiding the human race in metal-working may have been in Sheffield in Britain in 1856, in inspiring Sir Henry Bessimer's[346] nighttime dreams to come up with his process of blowing air through molten iron to produce steel.

What has happened is that a small group of miners, metallurgists and blacksmiths has taken civilization forward in leaps and bounds.

It is evident, though; that the Dwarfs consider their job mainly completed, and have withdrawn from many of their old bases. Still, it is interesting to see that the Dwarf colonies are near the centre of Italian steel-making, and near Sweden's furnaces. Only in Brazil and Argentina do they seem still to be active.

e)   THE GREEK CLASSICAL PERIOD

Between the dates 624 BCE to 270 BCE, the Greeks produced some of the greatest minds of the period:

- Hales (c. 624–546 BCE): Often considered the first philosopher, known for his astronomical observations and ideas about water as the fundamental substance.
- Anaximander (c. 610–546 BCE): A Milesian philosopher who proposed the apeiron (the indefinite or boundless) as the origin of all things.
- Pythagoras (c. 570–495 BCE): Known for his contributions to mathematics and his mystical philosophy centered on the transmigration of souls.
- Socrates (c. 470–399 BCE): Known for emphasizing virtue and the use of dialectic to pursue truth.
- Plato (c. 427–347 BCE): Socrates' student, famous for his theory of Forms and advocating for a polity governed by philosophers.

- Aristotle (384–322 BCE): Plato's most famous student, a polymath and founder of the Peripatetic school, whose writings span diverse subjects.
- Diogenes (c. 404–323 BCE): A leading figure in the development of Cynicism.
- Zeno (c. 334–262 BCE): Founder of Stoicism, a philosophy focused on achieving tranquility by accepting what one cannot control
- Pyrrho (c. 360–270 BCE): Credited as the founder of Skepticism and Pyrrhonism.

In part, this was due to Greece being a slave society, with Athens having about one third of its population as slaves. This gave the Citizens the time to indulge themselves.

It is not clear how much influence "Others" had in bringing this about and, if so, how did they do it?

Obviously this is a case where the skeptic is just as likely to be right as anyone else.

f)   CHINA

So far, I have been dealing with what is termed Western Civilization, but that is not the only one. Many human races' folklores tell of Gods visiting from the sky and teaching them the fundamentals of agriculture. Few of these progressed further.

The Mayans advanced into constructing massive stone buildings, but remained in the Stone Age.

The Chinese Bronze Age started in approximately 2000 BCE and, as the Chinese starting using bronze at roughly the same time as Britain, they could have been the recipients of knowledge from the same Anatolian source. They entered the Iron Age at roughly 500 BCE, again approximately matching the British timescale.

Chinese philosophers[347] emerged during 770–220 BCE, a period of intellectual flourishing known as the Hundred Schools of Thought, when they developed foundational concepts in Confucianism, Daoism, Mohism, and Legalism.

- Confucius (551–479 BCE),
- Laozi (c. 500 BCE),
- Mozi (c. 470–390 BCE),
- Mencius (372–289 BCE),
- Zhuangzi (c. 4th century BCE),
- Han Feizi (died 233 BCE).

These dates tie in with those of the Greek philosophers, so perhaps things were starting to develop in parallel.

This coincided with the rise of the first emperor of China, Qin Shi Huang (259–210 BCE), who unified China in 221 BCE. He established the Qin dynasty and is known for centralizing power, creating a unified script, building the Great Wall of China, and commissioning the Terracotta Army to protect his tomb in the afterlife.

There are claims that Qin Shi Huang was actually either an "Other" or assisted by "Others".[348] This would imply that

"Others" had decided to invest in a second civilization, just in case.

Here is another of those wonderful twists of fate. In the same period, both Greece and China went through a period of philosophical development, in one case protected by a slave society and in the other by a tyrannical ruler, as well as virtual slaves known as Coolies.

We are now into the stages of "Other's" syllabus where the telepathic species have to make a much larger contribution. I can only point out coincidences and possibilities. Proof is impossible, probably to the delight of our skeptical friend, but there is clearly a coincidence here.

Chapter 3

THE LAST THOUSAND YEARS

a) THE RENAISSANCE

From when the Romans left northern Europe, it descended into virtually continuous warfare for almost a thousand years. The Vikings invaded, the Saxons invaded, the Normans invaded, and Britain fighting France was almost a national sport, only recently replaced by rugby football.

When we finally came to the Renaissance, things settled down a bit, and we started the development of art, literature and philosophy. It began in north Italy, in the 14th century, and spread north, coming to an end in Italy with the Sack of Rome (1527), and elsewhere in the 17th Century.

In Italy, it was fashionable to sponsor artists such as Leonardo Da Vinci, Raphael or Michelangelo, so it would only have required "Others" to start such a fashion. Christopher Columbus managed to raise the funding for his expeditions.

The main cause of dissent in this period was the Roman Catholic Church, whose disapproval of the sciences caused Scientists to be labeled "Heretics" and, sometimes, as in the case of Bruno[349], this even resulted in their executions. Perhaps this was typical of human nature, to defend one's power, or perhaps it was instigated by a faction of "Others", who disagreed with the whole idea, fighting back.

The German Renaissance was inspired by the Italian Renaissance, with such famous personages as artists Dürer and Holbein, and the architects Wolf and Roritzer coming to the fore. Martin Luther had a major influence on the Protestant Reformation.

The Renaissance in Britain was inspired by the Italian and German Renaissances, with Shakespeare, Marlowe and Ben Johnson rising to prominence. The architects Inigo Jones and Christopher Wren flourished. King Henry VIII renounced the Catholic Church and became a Protestant. Explorers such as Drake and Raleigh obtained backing from Queen Elizabeth I. She survived a war with Spain and the Spanish Armada, but remained unmarried. The Renaissance in the UK was brought to an end by the English Civil War in 1642, when extreme Protestants took to smashing church sculptures which they considered idolatry.

There were certain key movers over this period with Martin Luther standing out. He changed northern Europe and, subsequently, North America for ever. What was his source of inspiration?

b)   THE RAILWAYS

The world's first steam train to run on rails[350] was built for a bet by Richard Trevithick in 1804 and successfully hauled a load of iron on the Penydarren Tram-road in Wales. However, this locomotive was too heavy for the tracks. The first practical steam train for public use was Locomotion No. 1, built by Robert Stephenson and Company for the Stockton and Darlington Railway, which opened in 1825 and carried passengers and freight.

The Rainhill Trials[351], held in October 1829 for the Liverpool and Manchester Railway, was the first significant competition for steam locomotives to determine the best motive power for the new railway. The L&MR directors had not decided whether to use fixed engines with ropes or steam locomotives and resolved to hold trials to see if a steam locomotive would meet their requirements. George Stephenson, Robert's father, who worked for the L&M surveying the proposed route entered the Rocket, designed by his son Robert, which emerged as the winner, completing the trials and establishing the steam locomotive as the preferred choice for railway transport.

What made all the entries, apart from Stephenson's Rocket, fail to complete the tests? Was this an example of "Other" intervention?

There followed what came to be known as Railway Mania, which was a speculative boom and subsequent crash in the 1840s, primarily between 1844 and 1846, characterized by a frenzy of investment in railway companies and rapid, often unprofitable, railway construction[352]. Following earlier successful lines, investors poured money into thousands of new railway projects, leading to inflated share prices, but by late 1846, the bubble burst, resulting in financial ruin for many and a slump in the market.

Thus the British railway system was built by people who subsequently went bankrupt, enabling a small group of companies to buy up the various lines, and to form a viable operation. All this expertise travelled round the world, uniting countries as never before.

Was Railway Mania something initiated by "Others" to make the Earth seem smaller?

Much of the Railway Mania was attributed to the so-called Railway King, George Hudson[353], who launched scheme after scheme, eventually owning over 1000 miles (1,600 Km) of lines. He could sell anything, but could manage nothing. Who gave him the drive to do this?

c) CHEMISTRY

In Europe and Russia, the early days of Chemistry were the realm of the individual with use of a laboratory of their own. Some of the most famous were:

- **Antoine Lavoisier**, (1743-1794) "The Father of Chemistry" His most famous discovery was the law of conservation of mass, which states that whilst substances may change their state or form, they retain the same mass. He was also the first person to identify oxygen and hydrogen
- **John Dalton** (1766-1844) is these days remembered for proposing his atomic theory, which represents the basics of chemistry today. He also studied and ultimately formulated a theory of atomic weight.

- **Amedeo Avogadro.** (1776-1856) His great discovery is what has come to be known as Avogadro's Law, which states that, under the same temperature and pressure conditions, the same volumes of different gases will have the same number of molecules.
- **Henry Cavendish**, (1731-1810) who discovered Hydrogen and proved that water was a compound, not an element, was intensely shy, and preferred to communicate by notes.

18th Century chemists received training primarily through apprenticeships, technical tours of mines and factories, and formal education at universities and specialized institutions like those for apothecaries.

The best training for a 19th-century chemist involved apprenticeships in busy shops or, increasingly, formal university education, with Germany eventually becoming the leading center for advanced chemical training by the mid-19th century. Early training relied on apprenticeships with a master, but this was often inadequate, leading to the development of specialized university laboratories and courses in Paris, Stockholm, and later, Giessen, Germany

So, why did Germany become the leading country for Chemistry?

18th Century Germany became prominent in chemistry training[354] due to strong state support, the development of a university-based research model emphasizing practical application, and the foundational contributions of prominent scientists like Stahl and Klaproth. This laid the groundwork for Germany's later dominance in chemical research and industry.

Apparently, the "Others" would only have to exert influence on a few policy makers to initiate Earth's development in Chemistry.

It is also recorded that August Kekulé[355] who in the 1860s first identified that the Benzine molecule was ring-shaped, and Dimitri Mendeleev[356] who presented the first periodic table of the elements in 1869, both said that they first thought of the idea in a dream. Did someone place it there?

d) ELECTRICAL POWER GENERATION & TRANSMISSION

Belgian Born Zénobe Gramme is credited with inventing the first commercially successful Direct Current generator, the "Gramme dynamo," in 1869. His 1871 working model, featuring a new wire-wrapped ring rotor known as the "Gramme ring," produced significantly higher voltages than previous designs and demonstrated that the device could also function as a motor. This invention spurred the development of electric power for industrial applications.

The "first design" for an electric generator was a simple homopolar generator built by Michael Faraday in 1831 to demonstrate electromagnetic induction. In the United States, Charles F. Brush developed one of the first efficient and reliable dynamos (D/C generators) in 1876, and Thomas Edison's Pearl Street Station in 1882 was the world's first purpose-built central power station, using his D/C system to power incandescent lamps.

The genius Nikola Tesla was born in Smiljan in Croatia in 1856. His father was a priest but he wanted to study the sciences. He was educated in the Real Gymnasium in Carlstadt, Croatia, before going to read engineering at Graz Polyrechnic and Prague University. His main interest was electricity and he had a flash of inspiration in which he visualized totally the workings of an alternating current (A/C) motor.[357] Did he jump or was he pushed?

He went to work for the Continental Edison Company of Paris, where he gained a reputation as a highly talented Electrical Engineer. Unfortunately he was very naïve when it came to business, feeling he had been cheated out of a promised bonus. In 1884, he moved to New York, where he was invited to work for Edison directly. He continued developing his A/C ideas in his own time, but received no encouragement from Edison. Once again he was promised a bonus for solving a particular problem, but never received it when he succeeded.

Tesla and Edison parted ways after Edison's failure to acknowledge Tesla's contribution – something Edison was known for. He then set up with George Westinghouse, and began a campaign to demonstrate the superiority of A/C. Edison responded by initiating a smear campaign to prove A/C was dangerous and even used A/C in the electric chair to create the perception that A/C was deadly.

Tesla and Westinghouse's demonstrations and the inherent advantages of A/C for large-scale power distribution led to its adoption. A/C power became the standard for the world's electrical grid, establishing the future of electricity and eclipsing Edison's direct current system, forcing him to wind up his company.

Tesla took out patents on his inventions, and these were subsequently bought by Westinghouse in 1888. However, Westinghouse proved as ruthless as Edison, paying for the patents, but not the subsequent commission which Tesla had expected.

Once again, Tesla had proved to be a brilliant engineer, but unworldly in business matters. We will come back to Nikola Tesla soon.

e) TELECOMMUNICATIONS

In 1831, the year Faraday invented the D/C electrical generator, an American, Samuel Morse[358], heard about it and thought that a method of sending messages along wires was needed. He was a Professor of Fine Arts at New York University, yet he was the person to dream up the famous Morse code. It is amazing what the "Others" can make you dream!

Italian Guglielmo Marconi's early experiments in radio communication were financed by the British Post Office, a significant relationship that involved the first open sea wireless transmission[359] in 1897 at Lavernock Point and the acquisition of Marconi's coast station technology in 1909. The Post Office, led by its Chief Engineer Sir William Preece, championed and supported Marconi's groundbreaking work in wireless telegraphy, which laid the foundation for modern radio and wireless communication systems.

However, Tesla had patented his Tesla Coil and radio-tuning device six years before Marconi[360]. In 1893 he gave a lecture on how a radio should work, and patented his system in 1897, but never challenged Marconi.

This is a classic example of the perfectionist and the pragmatist. Tesla could have patented his idea earlier, but he waited until he had got it the way he wanted. Marconi publicized each stage in the development of his product, catching the public's imagination and succeeding in raising the funding he needed. He was the one who got the Nobel Prize in 1909, and is remembered as the inventor of the radio.

f) CHINESE DEVELOPMENT

The Renaissance in Europe was preceded in China by almost a thousand years. The Tang (618–907) and Song (960–1279) dynasties are often considered analogous "golden ages" for China's significant advancements in culture, science, and technology.

Key Chinese inventions that changed the world include the "Four Great Inventions": paper, gunpowder, the compass, and printing (woodblock and movable type). Other significant innovations include silk, the abacus, the wheelbarrow, the seismograph, and the God's gift to mankind - tea. These inventions profoundly impacted fields such as navigation, warfare, communication, education, and trade, fostering cultural exchange and driving global development.

All these inventions worked their way into Western Europe, for good or for bad, driving its culture forward. In return, China gained railways, chemistry, electrical power and communications. China was not interested in many of the products of Europe's Renaissance, preferring to develop its own artistic style.

Whilst this growth proved beneficial to both cultures, it led to competition between them, which sadly remains with us.

Chapter 4

THE LAST CENTURY

a) <u>THE DEVELOPMENT OF US SOCIETY</u>

In the US, in the 18th century, a national elite began to grow. This was often based on inherited wealth being invested in expanding technology and a subjugated population of slaves and exploited immigrants. Some managed to break into this elite by hard work on their part, and even harder work by their employees.

It has been said that, once you have made your fortune, you can afford to have scruples, and some of the most ruthless families in making their riches are now the most famous of philanthropists. Just ask any steelworkers from the last two centuries.

The likes of Edison and Westinghouse had both proved to be ruthless businessmen, determined to suppress innovations where they would cost them, and unlikely to honor promises which would cost them, unless forced to. To some, this seemed the way to the top.

This elite worked together to further their interests. In the 19th century they had significant influence on the government and, by the 20th century, they effectively had behind-the-scenes control of the country.

Take the Bush family. President George H W Bush had previously served as Vice-President to President Ronald Regan, and before then he was Director of the CIA and Ambassador to the United Nations. His son, George W Bush also became President. His father was an investment banker and a Senator. His maternal Grandfather was an investment banker.

Or take the Kennedy family. Joe Kennedy Senior[361] was ambassador to the United Kingdom before World War II. One of his sons was President Jack Kennedy, and of his other sons Robert Kennedy was a Senator and became Attorney General, and Ted Kennedy was a Senator. His father P J Kennedy was also a Senator. One of his grandsons R F Kennedy became Secretary of Health and Human Services under President Trump.

Tesla was like a sheep amongst a pack of wolves from the moment he arrived in America.

b) POWER TRANSMISSION

Tesla was pretty wealthy once the battle of the currents was settled, although his future experimental projects were never going to be cheap. His first job was to oversee the construction of the Niagara Falls Power Station[362], and this enhanced his reputation. Then he struck out on his own, designing and patenting his Tesla Coil, and inventing and patenting radio years before Marconi.[363] He even built and publicly demonstrated a radio-controlled boat in 1897. He started working on the free distribution of wireless power.

In his notebooks from 1899, Tesla referred to 'stationary terrestrial waves' at approximately 7.5 Hz, a value strikingly similar to the Schumann Resonance that wasn't officially recognized until 1952. By energizing the earth-ionosphere cavity, he inadvertently tuned into the Earth's rhythmic pulse, years before scientific instruments validated its subtle presence.

Tesla needed money to develop his ideas, and he couldn't get a backer until, in 1902, J P Morgan[364], the driving force behind the General Electric Company (GEC), put up money in return for a 51% stake in the company holding the patents.

What Tesla didn't realize was that his wonderful ideas for free wireless distribution of power were a direct threat to GEC which had invested a fortune in their cable distribution system, and could wipe it out the way A/C had wiped out D/C and domestic gas.

At the first opportunity J P Morgan exercised his 51% controlling majority and, in 1903, stopped the work forever.

From then on, until his death in 1943, Tesla interested himself in many projects, was acclaimed as a genius, but faded into oblivion because his ideas seemed too fantastic at the time, even if they have often come to fruition much later, after his death.

He died in the Hotel New Yorker, and all his papers – 80 trunks full – were confiscated by the government. There were 163,911 documents yet Dr John G Trump from MIT, who was tasked to review them, was able to say - within 3 days – that there was nothing of military or security interest in them!

Yet when his papers were returned to his family, there were only 60 trunks full. The elite families had probably struck again – Dr John G Trump was the uncle of President Donald Trump. Tesla's ideas were not going to be used for the benefit of mankind but in the interests of the very rich minority, in a single country.

c)   EXTRA-TERRESTRIAL TECHNOLOGY

What the US elite most wanted were the power sources and the drive mechanisms of "Other" spacecraft. There were obvious differences between the drives of different species, as reported by witnesses[365]. Some gave off flames, and some gave off dangerous emissions, possibly radioactive or possibly electro-magnetic, which could cause burns, cancer or even death. It appeared that water was necessary for their operation.

Some "Others" had invisibility mechanisms, for their ships and portable. There are many descriptions of "Others" only visible from the waist up, or having no feet. This would be possible using such a device. Perhaps the portable power source for this device was limited. There are also reports of bullets passing through "Others" without harming them. Could this be the use of holographic imaging?

Many "Others" are described as levitating or hovering above the ground. This could be the result of portable technology, perhaps similar to the technology used in their spacecraft. Their non-lethal weaponry would also be useful, sending humans to sleep or controlling them.

Of course, the US elite were not the only ones after this technology. Russia and Nazi Germany were extremely interested as well.

Some of this technology would probably have been part of the "syllabus" of human education, but this had been interrupted.

d)   THE "OTHERS" TEAM

In Appendices 1 and 2, I have attempted to deduce which species were involved in the process of educating the Human race:

- **The Anunnaki**, possibly assisted by **Avians**, who first altered human genetics, and then taught them the basis of agriculture, mathematics, medicine, time measurement and many other things. They are reported to have moved on about 5,000 years ago.(See Appendix 3)
- **Dwarfs**, who possibly assisted humans in building monumental structures, and certainly taught them mining, and metal-working, both blacksmithing and metallurgy. They are reported to have withdrawn from human interaction in most places, and have returned to their earth colonies. (See Appendix 4)
- **The Pleiadians**, a telepathic tall Nordic race, who have probably been around as long as Dwarfs, and could have assisted in disseminating learning around the world.(See Appendix 5)
- **The Clarions**, a race of deceptively frail-looking humans with distinctive oriental almond-shaped eyes. These are telepaths who, because they did not stand out in a crowd, could have found it easier to approach those people who had to be influenced.(See Appendix 6)

- **The Tall Whites**, whose constellation of origin is unknown, but who probably oversee the whole project, and possible used their telepathy to influence human actions. They are still active all around the Earth. (See Appendix 7)
- **The Small Grays,** are responsible for defending Earth against attacks by malevolent "Others", and stopping anyone using atomic weapons. They also limited the power of the US elite, and later Majestic.(See Appendix 8)

This team had achieved so much before they came up against advanced human greed, which really put a spanner in the works.

e)   AN ALTERNATIVE SYLLABUS

If universal free power was the next step in this syllabus, it appears to have gone seriously astray. It would seem that the "Others" then revised their tactics.

They arranged a serious of UAP crashes around the world, including bodies. There were crashes in:

- Russia   1941   Rostov[366]
- Italy   1933   Magenta[367]
- The United States of America   1947   Roswell[368]

There were probably many more around the world, including India, China and the United Kingdom. This meant that, even if the US elite had sole use of Wireless Power at first, they could not prevent "Other" technology from being available worldwide.

The United States went all-out to get as many crashed UAPs as possible, using its diplomatic might and threats of force if necessary.

- There was even a tense confrontation with Canada[369] on 12th February 2023 in Lake Huron, where the Canadian government elected to back down. It later turned out to be a Chinese spy balloon.
- On 25th August 1974, north of Presidio in Mexico, it is alleged that an armed US recovery team crossed the border[370] and claimed that all the Mexican soldiers there were already dead. They took the UAP and felt they had to burn everything else, bodies included, to avoid contamination.

The United States has a long-established program of reverse engineering[371] UAPs, with the intention of producing their own versions, and Russia and China probably do as well.

f) THE MAJESTIC TWELVE

The Roswell Crash occurred in 1947. That same year President Harry Truman signed the National Security Act which set up the Department of Defense, the USAF and the CIA. It is claimed that an organisation called Majestic-12[372] was set up as the same time, reporting directly to the President, to deal with all matters relating to UAPs.

In 1949, Majestic's secretary, James Forrestal, died after falling from a 16th-floor window at Bethesda Naval Hospital. While officially ruled a suicide, his death was shrouded in controversy and sparked numerous conspiracy theories about a Majestic power-grab.

It should be noted that a number of the members of this organisation were members of the elite families so, once again, they have a firm grasp on the levers of power.

Within a very short time, Majestic managed to sideline even the President, and to set up almost limitless funding from the State via black budgets and from industry acting in its own self-interest. The military have always resented political interference, choosing to forget that they have sworn to protect the nation as defined by its elected government. Perhaps they have chosen to re-define what is meant by "nation".

In his farewell address when leaving the White House, President Dwight Eisenhower[373] said:

> **"In the councils of government, we must guard against the acquisition of unwarranted influence, whether sought or unsought, by the military-industrial complex."**

It is this military-industrial complex which has enabled Majestic to take control of the United States of America.

It is alleged that when President Carter took office, he was given an official briefing on UAPs and "Others" and that, afterwards he was found in his office, sobbing his eyes out[374].

The reverse engineering of UAPs has continued, probably now under the control of Majestic, and the US has now become a two part state. NASA undertakes the public part of the space programme, whilst other agencies finance the private sector to undertake Majestic's space programme. NASA has to be very careful to avoid accidentally releasing anything about Majestic's activities. In these areas, the President has to do what he or she is told.

It has been reported that Majestic's operations are now so advanced that they could leave Earth behind if they wished[375]. One wonders why some billionaires are so keen to develop their own spaceships, and go to Mars.

It certainly would not be in the US elite's interests to cooperate with other nations, although they might work with other ruling elites. Otherwise, it would imply that they are working in the interest of the Human Race and, as I've already shown, this is most unlikely to be true. They would have to have shown a considerable change of heart since they pushed Tesla aside. They are still ruthless in dealing with any challenge to their power.

g) CURRENT UNITED STATES ACTIVITIES

Another genius who may have been used by "Others" was Dr Werner von Braun. This German was instrumental in the development of the first cruise missile – the V1 – and the first ballistic missile to reach space – the V2 – before moving to the United States to lead their NASA space program developing the massive Saturn V rocket which took men to the moon.

He clearly believed in "Others", and possibly that is where his inspiration came from.

The US has achieved far more in space than has been admitted. They have autonomous vehicles such as the X-37B[376], and crewed triangles such as the TR3B[377]. The hacker Gary McKinnon[378] showed that they have a number of craft, paid for by the American Taxpayer, and there are probably more under construction.

The most recent American spacecraft[379], that they admit to, include the Orion capsule for deep space missions, the SpaceX Dragon spacecraft for low Earth orbit transportation, and SpaceX's large-scale Starship system designed for lunar missions and beyond. The private Blue Origin spacecraft, such as New Shepard, are used for suborbital tourism, while other private ventures like the Intuitive Machines are conducting lunar surface missions. Northrop Grumman's newest, large cargo variant, the Cygnus XL spacecraft, also recently launched on a mission to the ISS.

To achieve newer spacecraft, the US has had to set up its own bases, secret from the Earth population and, possibly, from the various "Other" groups.

There is certainly one US base in Australia. Pine Gap near Alice Springs is claimed by the US to be a listening station set up to overhear electronic communications primarily in China and North Korea. However, chance sightings demonstrate that it is far more than that[380].

The Blue Mountains in Australia are claimed to hold a second base used for building and testing US spacecraft. In my opinion the book which makes this claim[381] derives far too many conclusions from too little data, combined with hearsay. There have been many UAP sightings in the area, but have these been "Other" or US spacecraft, or both?

There has been one interesting possibility to have come from Majestic's probable usurpation of Tesla's early attempts at a wireless power project, and the later discovery of the Schumann Cavity, possibly building on his early work. On 8th July 2003, the US launched the Opportunity Rover to Mars, where it started operating in January 2004. It was designed to operate for 90 days. It now holds the record for the longest-lasting Mars rover, operating for almost 15 years (5,498 Earth days).

Before it was launched, there were some last minute alterations to the power supply system (Not in itself unusual). Could it be that they had discovered that there was a wireless power supply system operating on Mars, left over from some "Other" activities, and they fitted a power receiver to Opportunity?

The supposedly solar-powered rover's mission ended in 2018 after a global dust storm, with NASA officially declaring the mission complete in February 2019.

There are reports of an operating Wireless Power system on earth in an underground cavern under Mount Denali in Alaska, which is reputed to contain what is known as the Dark Pyramid[382]. It has been claimed that this generates sufficient electricity to power the whole of Canada[383]. Is this an ancient facility or has the US built it? How did the US find it?

Tesla lost his patents in 1903, and in 1908 the Tunguska Incident occurred, where something large apparently exploded in mid air over Russia. It has been suggested that it was shot down by copper "Cauldrons" spread out in Siberia, which are supposedly part of a defense system build by "Others" and powered by the

Dark Pyramid in Alaska[384]. Scientists in the US, looking at Tesla's patents, might have detected it powering up beforehand, and realized what was happening. However, it is probable that Human technology was not sufficiently advanced at that time, but there have been several objects seen to break up over Russia since then, so they've had several more chances.

There have been claims of an ancient pyramid[385] found on Mars, and a number of claims that the Great Pyramid at Giza[386] was once a power generator for the Ancients, using granite monoliths as relays.

The "Others" have succeeded in finessing the US elite, but it will be interesting to see what happens next.

Chapter 5

THE WAY FORWARD

a)   SPACEFLIGHT

You may remember that, back in Chapter 4, Section b, I described how it was found that the Schumann Cavity, the space between the ground and the ionosphere, could resonate so that it carried electrical power from the transmitter to anyone with the right receiver.

This may be how power is transmitted to all the "Other" craft flying over us, from Alaska and perhaps from as-yet undiscovered sites elsewhere. There would be no particular requirement for these craft to have inter-planetary capability, let alone inter-stellar. They could simply dock with mother-ships if this was necessary.

So, the various human nations have only been presented with examples of basically up-rated aircraft for them to reverse engineer. These do not use jet fuel and so are a step in the right direction. Also, they effectively negate the US's temporary lead which they achieved by confiscating Tesla's papers, and keeping them secret.

I suspect that the US's claim, that they could up-sticks and leave Earth if necessary, was just about correct. They could only fly beyond the ionosphere using rocket fuel powered craft, but these would be very slow when compared with "Other" inter-planetary ships.

The "Others" have managed to regain control of Human education. However, it may be some time before the next stage arrives.

b) TRADING WITH "OTHER" COLONIES

The various Earth nations would have had access to some of the "Other" technology on the crashed craft, such as their invisibility devices, and possibly their levitation devices, but the "Others" would have made sure there was no weaponry or mind-control equipment present.

However, most national governments would have some idea of where the "Other" colonies were in their country, even if they haven't told their general population. To avoid any confrontation due to resource limitation, they will probably enter into trade agreements with them. The colonies would still have the equipment they originally needed for construction, and they could either lease this out, or use it to undertake jobs for humans, in return for whatever supplies they needed.

It wouldn't be long before governments would be forced to admit what was going on, particularly if the "Others" announced that they had decided that "the time is right", as specified in the Small Gray's original contract[387] with the US.

It may be some time before the "Others" team decides to proceed to the next level in the "Syllabus". They would probably want to see all of this level's knowledge made available to the general population first.

What will come next?

THE UFO, E.T, ALIEN TRILOGY

APPENDICES

APPENDIX 1 - The Little People 249
i What are Little People? 251
    Table 1 Number of Little People in Argentina of different heights as estimated by witnesses 251
    Table 2 – Estimated heights of Little People in Argentina 253
ii Various Books' Descriptions of Little People 253
    Table 3 Various books' descriptions of "other" physiology 254
    Note: The book by Neil Anami does not have page numbers.

APPENDIX 2 - Taller "Others" 255
    Table 1 Number of Taller "Others" of different heights sighted in Argentina as estimated by witnesses. 257
    Table 2 Species taller than 1.5m 258

APPENDIX 3 - Anunnaki 261

APPENDIX 4 - Dwarfs 265
i Dwarfism 267
ii Mythology 268
iii The Dwarfs Today 269

APPENDIX 5 - Pleiadians 271

APPENDIX 6 - Clarians 275

APPENDIX 7 – Tall Whites 279

APPENDIX 8 - Small Grays 283

APPENDIX 1
The Little People

THE UFO, E.T, ALIEN TRILOGY

THE LITTLE PEOPLE

i. WHAT ARE LITTLE PEOPLE?

When someone is called small, what does that really mean? It could mean absolutely tiny, like a new-born babe, or it could mean a bit smaller than the speaker. This gives a range of about 0.3m (1') up to say 1.6m (5'3").

In an effort to understand how people think, I have looked at the complete list of "Other" sightings for Argentina from the book published by George Mitrovic[388], extracting those sightings of entities 1.4m or less in height. I chose this upper limit to avoid confusion with Small Grays. I have chosen Argentina because North America is dominated by Small Grays and Bigfoot (Yeti), and Europe seems to have a lot of different types of "Other" Little People.

Short	<0.5m	0.5m	0.6m	0.7m	0.8m	0.9m	1.0m	1.1m	1.2m	1.3m	1.4m
77	16	9	19	17	3	27	50	4	23	4	13

Table 1 Number of Little People in Argentina of different heights as estimated by witnesses.

Very small "Others" really only started being sighted in about 1965, and Small Grays, normally described as being 1.4m – 1.6m, only started abducting humans in South America on a big scale in about 1975.

It would appear that many people have difficulty in estimating heights without some sort of reference, whether their own height, or some convenient object. It is evident that over a third of the witnesses preferred to avoid the issue. So, some of these sightings of short people could be for entities up to 1.6m high, so I have had to exclude them.

Amongst the sightings, a half of those at 1.0m were green-skinned, and most at 0.6m were described as Cyclops. There were many sightings of really tiny Little People, less than 0.5m tall.

Assuming that all the real Little People are within the range 0.1m to 1.4m, I have looked at the various descriptions given in the sighting reports, trying to match them with their estimated heights where possible.

In Table 2 on the next page, I have tried to match the sightings with known species of "Others" where their height range fits. In Table 3, I have surveyed the various publications which attempt to describe these "Others", in search of commonality. Again, I have only considered those species described as shorter that 1.5m.

Together, Table 2 and Table 3 demonstrate a wide range of physiological variations in the Little People, and this implies that there are different species present rather than slight variations in one species. On this basis, there would appear to be no real reason why Dwarfs should not be considered as one more member of the Little People range of "Other" species, albeit about half with green skins.

Not all of these "Others" arrived on Earth at the same time. It is claimed that the species, called the Chaneques by the Mexicans, are the descendants of a spacecraft which crashed there thousands of years ago[389], whilst the 7 cm tall species, called the Jenglot by the Malaysians and Indonesians, only arrived in their spacecraft in the 1960s[390].

Bearded Brown hair brown or green skin	1.0m – 1.3m	Dwarf	
Bearded Red hair	1.0m – 1.3m	Dwarf	
Gray/Green Gnome	0.9m		
Manlike	0.2m	Chaneques	Mexico
Tiny Goggle Eyed	<0.1m	Jenglot	Malaysia Java
3 Horned	0.4m		
White Skin, Wings	1.0m	Sprite	
Young Boys/Girls			
Long Eared	0.6m		
Skin Like Acne	1.4m		
Triangular Body			
Large Headed	0.5m – 1.3m		
Green Horned	0.5m – 1.4m		
Square Headed		Ant People	
Hairy	0.7m		
Bald headed	1.2m		
Fangs & Claws	1.0m	Chupacabara	
Gray Large head & eyes	1.2m – 1.5m	Small Gray	

Table 2 – Estimated heights of Little People in Argentina

ii. VARIOUS BOOKS' DESCRIPTIONS OF LITTLE PEOPLE

This has proved an eye-opener. Out of 8 books, they can seldom reach agreement on any "Other" apart from the Small Grays, and even then it is only 5 out of 8 who do. Only Dwarfs and Arcturians get mentions by 3 out of 8 books, and then we are down to 3 species getting mentioned by 2 out of 8 books, and 9 species getting a single mention.

Name	Height (m)	Pastore[391]	Spartacus[392]	Fredrich[393]	Anami[394]	McDaniel[395]	KGB[396]	Campobasso[397]	Sokol[398]	Details	Planet
Small Gray	0,9-1.3	77	4	-	?	-	-	157	63	Big Eyes	
Dwarf	1.0-1.3	-	-	33	?	27	-	-	-	Hairy	
Arcturians	1.0-1.3	-	74	19	-	-	-	-	14	Blue/green	
Langs	<0.70	258	-	-	-	-	45	-	-	Fairies	
Iguanoids	<1.0	-	-	-	?	-	-	231	-	Iguana	
Spotted Face	<1.0	-	-	-	-	55	-	258	-	Acne	
Sprite	<1.0	-	-	-	-	53	-	-	-	Wings	
Sylphans	1.3-2.0	-	-	-	-	-	-	-	156	Wings	
Batbazouls	0.7-1.3							257		Wings Ugly	
Goblin	<0.9	-	-	-	-	35	-	-	-	Big Ears	
Zeta Reticulan	1.3-1.9	-	-	-	-	-	-	175	229	Head & Eyes	
Little Green Men	<1.0	-	-	-	?	-	-	-	-	Green	
Ellina	Small	-	-	-	-	-	76	-	-	Elvish	
Dorsay	<0.5	-	-	-	-	-	39	-	-		
Small Reptoid	<1.0	-	-	-	-	-	-	253	-	Raptor-like	
Zeta Lizard Human	1.0-1.7	-	-	-	-	-	-	211	-	Large Head	

Table 3 Various books descriptions of "other" physiology
Note: The book by Neil Anami does not have page numbers.

It would appear that the Little People are of much less interest than "Others" that look like humans or are far, far taller. Nevertheless, it seems clear that Dwarfs and Arcturians seem the most common apart from Small Grays.

APPENDIX 2

TALLER "OTHERS"

THE UFO, E.T, ALIEN TRILOGY

TALLER "OTHERS"

Once again, I have tried to estimate the most likely of the taller "Others" to be present in Argentina[399] by looking at the frequency with which they have been sighted. There is a very wide range in heights, even when the Little People are excluded. There are a large number of sightings at 1.5m because this includes the Small Grays. On top of the sightings recorded here, there were 16 Yeti, 8 Cyclops, 7 Reptilians and 4 black Mothmen.

1.5m	1.6m	1.7m	1.8m	1.9m	2.0m	2.1m	2.2m	2.3m	2.4m
21	13	41	22	10	43	18	26	30	26

Table 1 Number of Taller "Others" of different heights sighted in Argentina as estimated by witnesses.

It appears that there are 3 peaks, at 1.7m, 2.0m and 2.3m. Hopefully these will coincide with the 3 species which are in the majority.

Looking at the same books which I used in Table 3 of Appendix 1, to identify the various species of Little People, I have listed, in Table 2 below, all the "Others" described, which are not shown as Little People. The names seem to be quite arbitrary – Take any star or creature and make it into an adjective. The KGB list seems only to have Small Grays in common with this list. Some descriptions are flatly contradictory.

With so many "Others" being described as tall Nordics or tall and human-like, you are left wondering how we ever knew that these were different species.

Name	Height	Telepath	Description
Agathans	2.0-2.6m	N	Earth human (Telosians)
Alcyone Pleiadians	1.6-1.9m	Y	Nordic
Altarians	2.0-3.0m	Y	Human-like Blue/green/tan skin
Andromedans	1.7-2.1m	N	Blue skinned
Antarians	2.2-2.8m	Y	Human-like
Apunians	2.1-2.8m	Y	Nordic
Arcturians	1.3-1.6m	Y	Blue skin
	3.0-4.0m		
Arians	1.7-2.1m	Headgear	Nordic
Cassiopians	1.9-2.5m	Y	Human-like. Webbed fingers
Certans	1.9-2.5M	N	Human-like
Clarions	1.5-1.7m	Y	Petite human-like, slanted almond eyes
Cyclops	1.9-2.0m	Y	One eye
Cygnus Alpha	2.2-3.0m	Y	Tall human-like
Eridaneans	2.0-2.2m	Y	Nordic. Light blue skin
Itipurians	1.8-2.0	N	Thin humans
Lyrians	2.0-3.0m	Y	Feline
Klermers	2.5-3.2m	Y	Tall human-like
Koldashans	1.7-2.0m	Y	Human-like
Lady of Light	1.6m	N	Fair human. Skin glows
Lyrans	2.0-3.0m	Y	Human-like
Mantis	2.3-3.2m	Y	Insect. Triangular head
Melchizedeks	1.7-2.0m	N	Human-like
Mothman	2.0-2.6m	N	Dark fur, big wings.
Original Human Orions	3.3-3.8m	Y	Human-like
Pleidians	1.8-2.1m	Y	Nordic Fair graceful, slender
Procyonians	2.0-2.2m	Y	Feline
	1.8-2.8m	N	Human-like
Proxima Centaurians	1.6-1.7m	N	Human-like
Renegade Pleidians	2.6-4.2m	Y	Massive human-like
Reptoids	2.0-3.0m	Y	Scaly skin
Sagittarians	2.5-4.0m	Y	Human-like

Name	Height	Telepath	Description
Saurians	1.8-2.2m	N	Lizard-like with tail
Sirians	2.0-2.7m	Y	Blue skin gills webbed hands
Soulzars	3.0-5.0m	Y	Hairless hybrids. Bald
Sylphons	1.2-1.8m	N	Gossamer wings
Tall Whites	2.1-3.1m	Y	Slender body
Thurbans	2.1-2.7m	Y	Reptilian
Titan Sirians	2.5-4.0m	Y	Blue skin, elongated head pointed chin
Unmites	2.5-2.8m	Y	Nordics
Vegans	1.8-2.2m	Y	Human-like
Venusians	2.0-2.8m	Y	Elegant, glowing skin (Valiant Thor)
	1.8-2.1m	Y	Human-like
Zeta Reticulans	1.0-1.6m	Y	Fragile body, big head & Eyes (Grays)

Table 2 Species taller than 1.5m

Based on these, the 1.7m peak could be caused by about 7 species but, only 3 of these are telepathic and many parts of the "syllabus" need a telepathic ability. This leaves Andromedans, Clarions and Melchizedeks. Clarions[400] are reported to have slanted oriental eyes, and these have been described in many places around the world, so these are probably the best bet.(eg Switzerland 1988)[401]

The wider peak between 1.9m and 2.2m points towards a Nordic of one form or another. The type whose height range best fits is the entity called a Pleiadian, and they are also telepathic. These are the claimed ancestors of many islanders in the Pacific Ocean as well as some Central American tribes.

The third broad peak between 2.3m and over 2.5m seems to belong to the Tall Whites, who are telepaths too.

THE UFO, E.T, ALIEN TRILOGY

APPENDIX 3

Anunnaki

THE UFO, E.T, ALIEN TRILOGY

THE ANUNNAKI

The Anunnaki[402] are reputedly the "Other" species who first colonized Earth, perhaps over 50,000 years ago. They come from the planet Nibiru. It is claimed by some that this planet orbits our Sun on a vast elliptical orbit so that its appearances are very rare – every 3,600 years. However, I suspect that our astronomers would have found it by now, if that were the case.

They are described as very tall (perhaps 4-5 meters) humanoids, with the males having long beards, and they are often depicted in carvings as having wings.

They are claimed by Maestà Pastore[403] to have come to Earth in search of gold, supposedly to seed their atmosphere to create a greenhouse effect to keep their planet warm when it is distant from our sun. However, they decided to create a slave race to do the hard work.

Apparently speed was not of the essence, so they decided to modify the genes of a large primate, and then they sat back to wait for us to develop. I doubt if they remained on Earth whilst this had its effect, but they returned in about 10,000 BCE, to turn us into a self-sustaining workforce.

We are essentially someone's property.

They set up a colony in Sumeria, and started to teach our ancestors the basics of agriculture, medicine, animal husbandry, mathematics and astronomy. They must also have done this in the west of Central and South America, and in South Africa because these are areas where gold is plentiful.

In his book, Pastore gives an account of the Anunnaki politics which led to their colony being abandoned about 5,000 years ago, and we humans being apparently left to our own devices.

APPENDIX 4

Dwarfs

THE UFO, E.T, ALIEN TRILOGY

DWARFS

i. Dwarfism

There are about 200 types of the condition known as dwarfism, and the birth rate in the United Kingdom is about 1 in 25,000[404]. In the past, without proper pre-natal care, this rate would have been much higher.

Of these births, 70% are what is termed "Achondroplasmia". There are 3 main visible types:
- Proportionate dwarfism where the individual is small but otherwise is perfectly normal.
- Disproportionate dwarfism where the longer bones of the body, particularly the legs, are much shorter and may the deformed.
- Seriously deformed people, perhaps even missing limbs, and often with mental deficiencies.

Back in history, these small people had to trade on their deformities if they could. There were troops of acrobats in Roman times, they worked as "Fools" to entertain the rich in the Middle Ages, and sometimes became their trusted advisors[405]. It has been wrongly suggested that people suffering from gigantism are more stupid than average. What is probably true is that people suffering from dwarfism seem to have more wit about them than average, given the proportion of them that rose to important positions. For example:

- Jeffrey Hudson, a small person born in 1619, became a trusted messenger for King Charles, and was eventually knighted.

- The famous Italian small person Bertholde became Prime Minister to the King of Lombardy.

What is missing is that you do not hear, today, of such small people being famed for their stone-working and metal-working skills, yet that is the tradition which is held in the folklore of many nations.

ii. <u>The Mythology</u>

There is a clear distinction between the Small People who suffer from dwarfism, and the Dwarfs of mythology and extra-terrestrial visitations, who are a separate race, perhaps of "Others"

In his book, Claude Lecouteux[406] looks at the folklore of Europe, particularly Germany, France and Norway. He notes, in particular, that Dwarfs have been in northern Europe long enough to have joined the fringes of their pantheons, but that they appeared to have faded away in Victorian times, leaving the lore of Dwarfs working away in their secret smithys.

He describes how, in pre-Christian times, they were considered generally friendly, but the Church gradually demonized them:
- In France, Aubéron, who is described as King of the Dwarfs and a great magician, is noble and knightly in the manner of the Knights of the Round Table.
- In Germany, Alberich was father of the King of Lombardy and helps him win his bride. He may be model on which the story of Aubéron is based. However, he is also described as a masterthief.
- In Denmark, Dwarfs ruled the land of the dead.

- In Norway, the Dwarfs forged magical artifacts, and possessed hoards of gold

In Norse mythology Odin[407] the one-eyed Dwarf is the All-Father, the god of wisdom, war, magic, and death. He sacrificed his eye to drink from the well of Mimir, gaining vast knowledge. He also hanged himself from the World Tree, Yggdrasil, to learn the secrets of the runes. He is often depicted as a one-eyed, long-bearded old man.

In Spain there is the Duende[408], a mischievous spirit, similar to an elf or goblin, that lives in people's homes or in the wilderness. It can be a fierce protector of its dwelling and may bring good or bad luck. A benevolent Duende might perform helpful tasks or leave small gifts, while a displeased one could play pranks or cause disturbances.

The Mayans have a myth about a Dwarf[409]. He became king, then built a palace, and then a city which is part of the Uxmal ruins.

iii. Dwarfs Today

Dwarfs' fame as miners comes from their diaspora when the Mediterranean tin became scarce and they spread out to find more, at the same time teaching the arts of mining and metal processing.

Since then, the mythology of Dwarfs appears to have faded, although their physical presence is still reported occasionally in sightings around the world, with evidence of at least two colonies[410], one in Northern Italy and one somewhere in Sweden. There are numerous sightings in Brazil and Argentina, suggesting that there is another one there, perhaps near Cape San Roque.

Anatolia, the source of early metal technology has no records or tales of Dwarf sightings in living memory[411].

APPENDIX 5

Pleiadians

THE UFO, E.T, ALIEN TRILOGY

THE PLEIADIANS[412]

As you might imagine, Pleiadians come from the Pleiades star cluster. A number of different "Other" species[413] are reputed to come from this cluster:

- **Renegade Pleiadians**. These are the tallest of the Pleiadians at 2.1m – 4.0m. They are greedy and malevolent, being basically self-centred.
- **Alcyone Pleiadians**. These are slightly shorter than the Pleiadians at 1.7m – 1.9m. They are generally kindly and friendly
- **Pleiadians**. These are typically humanoids who are slightly taller that humans (about 2m) and are muscular, fair skinned and blond haired[414]. For this reason they are also described as "Nordic".

They are fully telepathic, capable of conveying their thoughts amongst themselves, and with other species, whether they are telepathic or not.

Various cultures have mythical figures or groups of beings they associate with the Pleiades star cluster[415]:

- various Native American groups like the Kiowa and Nez Perce,
- in the Andes, where the Pleiades were seen as a symbol of abundance and were associated with the harvest season,

- in Hinduism where the Krittika group is identified with the Pleiades and
- in Hawaii, New Zealand (Māori), the Cook Islands, and Tuamotu,

Across Polynesia and beyond, the cultural significance of the Pleiades extends beyond a single island, serving as a celestial guide for navigation, agriculture, and ceremony.

Clearly, in the past, they were one of the groups who sought to teach races the first steps in civilization.

Even today, they often directly assist the more primitive tribes by offering healing, although they prefer not to become involved more than this.

APPENDIX 6

Clarions

THE UFO, E.T, ALIEN TRILOGY

CLARIONS[416]

Clarions are a humanoid race indistinguishable from humans, although they do have one marked characteristic – slanted almond shaped eyes that look vaguely oriental. They have petit facial features, but they do not stand out in a crowd, making them very useful in passing themselves as human.

However, there have been a number of occasions when a witness has commented on their eyes[417], suggesting that they are very much involved in further the "Other" syllabus. They are also telepathic, so they have been important influencers for many years.

Little is known about their way of life other than that they are a human extraterrestrial race, originating from the Aquila constellation, near a binary star. Some have been on Earth for 80 years, and is said that one of their bases is in the Amazon jungle.

They use disk-type UAPs, where the skin can become transparent, and it is said the ship is a living metallic creature.

Clarions first came to public awareness when they were implicated in a doomsday prediction for December 21st 1954 in the USA. Dorothy Martin[418] claimed to have received psychic messages from a planet called Clarion predicting flooding over much of North America and Europe. Some of the believers took significant actions that indicated a high degree of commitment to the prophecy. Some left or lost their jobs, neglected or ended their studies, ended relationships and friendships with non-believers, gave away money and/or disposed of possessions to prepare for their departure on a flying saucer, which they believed would rescue them and others in advance of the flood. I suppose they are still waiting.

We have no real evidence, apart from the word of Dorothy Martin, that Clarion was involved. I cannot quite see how this would have advanced the "Other" syllabus if they were, so I doubt that they had anything to do with it.

APPENDIX 7

Tall Whites

THE UFO, E.T, ALIEN TRILOGY

TALL WHITES[419].

There is some doubt about where these "Others" come from. Some say they are from Betelgeuse and are refugees on Earth[420], others say Antarius[421] or from nearby

They are certainly tall, with heights ranging from 2m to 3m, with pale white skin, almond-shaped eyes, and are like tall thin humans. Their features are thin and angular, and they are very graceful. Despite their thin physique, they are very strong. They are telepathic, highly intelligent and live for centuries.

Their spacecraft use advanced anti-gravity propulsion and inter-dimensional travel, letting them move vast distances almost instantaneously.

They often carry devices which can manipulate electro-magnetic fields, and which can be used for personal protection. They are very skilled in healing.

They are generally thought to be here to monitor our technological advancement, and are considered by "Others" races to be capable mediators. They are reported to have had regular interactions with humans in a military and scientific context.

They have been on Earth for centuries, influencing the development of human cultures.

APPENDIX 8

Small Grays

THE UFO, E.T, ALIEN TRILOGY

SMALL GRAYS[422]

These are the most easily identifiable of the "Other" species on Earth, being 1m - 1.5m tall with, as you might imagine, gray skin. They have very large heads, pointed chins, a small mouth and nose, and very big black eyes which dominate their faces. They have long thin arms and legs, and their hands can have 3 or 4 digits.

They are reported to come from Zeta Reticuli, and it is thought that they were a biological race, which has devolved into a part synthetic form. It is suggested that the genetic experiments, which they perform on humans, are directed towards recovering their lost biological selves.

They offered to assist the US government after a threatening show of force over Washington in 1952[423], by some unknown species – probably Small Grays! The entered into an agreement with the US, whereby they offered technology, in return for a base in the US, and permission to take animals and humans for study. This was agreed provided the number of humans was limited and they were returned unharmed.

They are reported to be unemotional and calculatingly cynical, and it wasn't long before they were ignoring the agreement when it suited them. They have been on earth for some time, and have managed an effective take-over of the United States, In particular, they are now responsible for most of the human abductions, performing operations on humans, particularly females, and mutilating animals. They initially acted solely in the US, but their activities are now spreading world-wide, with abductions reported in South America, North Africa, Pacific Islands and even New Zealand.

They are reputed to have caused the US, Russia and the UK to be unable to keep their nuclear weapons at readiness, by shutting down missile silos, and shooting down test missiles.

However, they are also responsible for protecting Earth from attack by hostile "Others". This is probably just a spin-off from protecting the "Other" colonies on Earth, but we humans are along for the ride.

They have the agreed US base in Dulce, New Mexico, another near Catalina Island on the west coast of the US, one in Scandinavia and one somewhere in Europe.

Their abilities include telepathic control of humans, which facilitates their abductions.

PART 4

INDEX
&
REFERENCES

INDEX

A

A/C 227, 233
Abacus.. 229
Abductions 216, 285, 286
Admiral Byrd..............................22, 63
Agriculture 213, 218, 235, 263, 274
Air-Traffic Controller 23
Akhenaten................................13, 15
Alabrans 91
Alaska 17, 27, 64, 65, 68, 79, 241, 242, 243
Alberich... 268
Albert Coe 49
Aleutian Islands..........................64, 68
Algarve....................................131, 195
Almond-Shaped Eyes 235, 277, 281
Amphibian...................................... 97
Anannaki 87, 102
Anatolia 215, 219, 270
Anaximander 217
Ancient Mars Pyramid.................. 242
Andros Island 127
Ant People 58
Antarctica 22, 63, 69
Anti-Gravity 281
Anubis (Dog)................................. 151
Anunnaki ... 15, 77, 102, 149, 156, 209, 211, 212, 213, 235, 248, 261, 263, 264
Anzu 119, 149
Apennines..................................... 139
Apunians.. 146
Arcturians.......130, 145, 146, 157, 253, 254, 258

Argentina............ 38, 39, 40, 65, 67, 69
Aristotle..218
Arizona...................................10, 62
Artefacts 9
Arthur C Clarke 53
Asteroid 14
Astronauts 28
Asuras..............................16
Atlit-Yam 59
Aubéron..268
Australia .. 3, 10, 17, 41, 42, 44, 66, 68, 69
Avians 15, 102, 103, 119, 149, 159, 176, 212, 235
Avogadro..225

B

Bahamas 67
Baltic 19, 133, 136, 157, 180, 197
Barcelona.......................................216
Base..22, 23, 24, 26, 30, 32, 33, 37, 39, 40, 41, 42, 46, 47, 51, 54, 61, 63, 65, 66, 67, 72, 80, 207, 217, 240
Basque..216
Belgian Wave 11
Ben Johnson 222
Benzine..226
Bermuda Triangle.. 121, 122, 126, 161, 188
Bertholde 268
Bessimer..216
Bigfoot 56, 216, 251
Black Budgets 238
Black Forest 20

Black Sea 18, 30, 91, 136, 157, 180, 181
Blacksmiths 217
Blue Mountains.........51, 116, 174, 240
Blue Origin 240
Bob Lazar 25, 61
Boomerang 12
Bound Head 15
Brain Development 210
Brazil 39, 40, 69
Broadhaven.............................. 10, 32
Bronze 215, 219
Bronze Age............................215, 219
Bruno 222
Buenos Aires 67, 69

C

Canine........................... 115, 150, 151
Cape San Roque........................... 269
Cappadocia 18, 59
Caribbean..84, 126, 127, 152, 157, 161
Cartagena............................. 129, 187
Cassiopeians (Web-fingers)... 105, 146
Catalina 27, 67, 80, 108, 148, 154, 166, 167, 169, 286
Cauldrons........................17, 79, 241
Cavendish 225
Caves 88
Chaneques......110, 119, 124, 148, 252, 253
Chemistry 205, 224, 225, 226, 229
Chernobyl 30, 77
China 12, 16, 17, 46, 48, 67
Christopher Columbus 221
Christopher Wren 222

Chupacabra.. 37, 56, 83, 108, 112, 151, 165, 171, 188
Church 222, 268
Circular Mass of Reeds.................. 43
Clarions 235, 258, 259, 275, 277
Clay Tablets................................. 209
Cloud Formations 209
Colonies... 70, 77, 85, 92, 99, 100, 104, 105, 209, 110, 111, 112, 113, 118, 119, 120, 121, 124, 126, 127, 128, 129, 130, 131, 133, 135, 137, 139, 142, 146, 147, 148, 149, 150, 153, 154, 156, 157, 159, 207, 217, 235, 244, 269
Communications.....................229, 240
Compass 229
Confucius...................................... 219
Copper 215
Cosmonauts 31
Crashes 236
Cryptids 66
Cuba 67
Curiosity 215
Cyclops.. 108, 111, 135, 145, 164, 169, 198, 252, 257, 258

D

D allele 211
D/C 226, 228, 233
Dagon 90, 156
Dalnegorsk.................................... 30
Dalton 224
Damascus steel 215, 216
Daniel Fry...................................... 50
Dark Pyramid 241, 242
Darlington, Perth 45

Deep Cave 58
Denisovians 211
Devas... 16
Die Glocke..................................... 25
Dinosaur 14, 53, 54, 58, 93
Diogenes....................................... 218
Dr John G Trump...................233, 234
Dr Werner von Braun 239
Dragon's Triangle............. 3, 41, 42, 68
Drake 222
Drone.................................12, 22, 49, 80
Duende 269
Dugway Proving Ground................. 62
Dulce 26, 66, 68, 107, 154, 286
Dürer 222
Dwarf 39, 103, 111, 114, 117, 119, 122, 126, 127, 128, 130, 135, 138, 139, 147, 148, 149, 150, 152, 152, 156, 157, 169, 174, 180, 181, 187, 191, 198, 202
Dwarfism 267
Dwarka 59

E

Earthquake......................138, 148, 210
Edison.....................226, 227, 228, 231
Edwards Air Force........................... 24
Egyptian 13, 15, 213
Electrical Power 229, 243
Elite 231, 232, 234, 236, 238, 239
English Civil War 222
Enoch... 16
Espora Air Base............................... 40
Explosion....... 108, 122, 138, 148, 166, 180, 183
Extra-Terrestrial 205, 209

F

Falkirk Triangle....................32, 65, 69
Fantasy 210
Faraday...228
Feline..150
First Ballistic Missile.....................239
First Cruise Missile 239
Floating Rock 15
Flores..211
Flying Saucer 21, 42, 49, 64
Folklore 91, 99, 110, 111, 112, 124, 132, 148, 210, 212, 268
Foo Fighters 21
Fossil 14, 58
Frog..111, 185
Fukushima..............................118, 149

G

Gary McKinnon..............................239
Gateway 57
Gateways 5, 37, 60
GEC..233
Genetic Anomaly................ 14, 88, 102
Genetic Mutations 20
Genoa 139, 147, 157
Genome 95
Geoglyphs 137
George Adamski..............................50
George Hudson..............................224
George Stephenson.......................223
George W Bush 232
Ghost Fliers 133, 135, 197, 198
Giza 18
Globsters 115, 142
Goatsuckers 151

Göbekli Tepe 18
Gold 46, 60
Granite 15, 18
Great Flood 18, 59
Great Pyramid at Giza 242
Greeks 213, 217
Green-Skinned 252
Guantanamo Bay 127, 188
Gunpowder 229

H

Hacker 75
Hales 217
Han Feizi 219
Hawaii 274
Heretics 222
Hessdalen 132, 197
Hill 611 30
Holbein 222
Holographic 234
Hominid 211
Horseshoe Lagoon 42
Horus 119, 149
Hotel New Yorker 233
Howard Menger 51
Hudson Valley 11, 26, 62
Hundred Schools of Thought 219
Hypnotic Regression 109

I

Ice-Age 18, 58, 87, 88, 91, 99, 105, 115, 121, 143, 156
India 16, 59, 64
Inigo Jones 222
Intra-Terrestrials 4, 55, 58

Intuitive Machines 240
Invisibility 234, 244
Ionosphere 243, 244
Iron 215, 216, 223
Iron Age 215, 219

J

J P Morgan 233
J Rod 25, 62
Jack Picket 11
Japan 3, 15, 16, 41, 42, 68, 69
Japanese Airlines flight 1628 27
Jeffrey Hudson 267
Jenglot 252, 253
Joe Kennedy Senior 232

K

Kaikōura lights 44
Kalahari Desert 48
Kamchatka 105, 163
Kecksburg 25
Kekulé 226
Khufu 18
Kingman 25, 61
Klaproth 225
Koala Peninsular 30, 69
Kurushekta War 16

L

Laguna Cartagena 37
Lake Baikal .. 67, 69, 91, 136, 137, 146, 157, 199, 200
Lake Cartagena 126
Lake Chapala 109, 148, 157

Lake Erie 216
Lake Fluxian 67
Lake Huron 237
Lake Storsjon 134
Lake Titicaca 110, 111, 169
Laozi 219
Lapland 134, 135, 198
Lavoisier 224
Lemuria 92
Lenticular 9
Leonardo Da Vinci 221
Levitation 234, 244
Little Green Men 147
Little People .. 103, 110, 113, 117, 121, 124, 127, 130, 135, 148, 149, 152, 156, 157, 248, 249, 251, 252, 253, 254, 257
Lt. Col. Charles Halt 33

M

Magenta 236
Mahabalipuram 59
Majestic ... 71, 72, 73, 74, 75, 205, 208, 236, 237, 238, 239, 241
Malmstrom Air Force Base 27
Malta 15
Mantis 107, 109, 165
Mantras 5, 56
Mar del Plata 130, 158, 191, 192
Marconi 228, 232
Marlowe 222
Mars 239, 241, 242
Marsh Gas 210
Martin Luther 222
Mayans 213, 218, 269
Mediterranean 91, 156

Megaplatanos 35
Men in Black 12, 22
Mencius 219
Mendeleev 226
Menhirs 18
Mesolithic 15
Mesopotamia 212
Metallurgists 217
Meteor 210
Meteorite 9, 20
Meteoritic Iron 215
Mexico .. 16, 18, 26, 37, 38, 64, 67, 68, 69
Michelangelo 221
Microcephalin Gene 211
Military-Industrial Complex 238
Milwaukee Deep 37
Mind-Control 244
Miners 215, 217
Missile Test 26
Mohenjo Daro 16
Moon 18, 63
Morse 228
Mother-Ships 11, 243
Mothman 130, 149, 257
Mount Denali 64, 65, 68, 79, 241
Mount Kailash 64, 69
Mount Musiné 64, 65, 69
Mount Popocatépetl 64
Mount Shasta 64, 65, 68
Mount Shishaldin 64, 68
Mountain 16, 48, 59, 64, 6888, 100, 132, 159, 171
Mozi 219
Multiple Universes 5, 55
Musinè 137, 138, 147, 201
Mutilations 26, 47, 66

N

Nan Madol 15, 42, 118, 213
NASA 74, 75, 76, 238
Neanderthals 211
Nests 43
New Zealand 3, 42, 44
Niagara Falls 232
Nibiru 15, 263
Nobel Prize 229
Nommo 90, 156
Nordics 24, 50, 51, 72, 77, 105, 112, 117, 121, 126, 127, 128, 130, 146, 152, 157, 171, 176, 181, 187, 191, 192, 257, 259
Northrop Grumman 240
Nuclear .. 16, 20, 21, 23, 24, 27, 31, 32, 42, 50, 51, 57, 72, 77
Nuremberg 9, 19, 79

O

Oahu 120, 181
Oannes 90, 102, 156
Obelisks 17
Octopus 95, 142
Odin 269
Officialdom 210
Olympians 16
Operation High Jump 22, 63
Operation Mainbrace 21, 22
Opportunity Rover 241
Orange ... 129, 130, 139, 145, 165, 166, 167, 168, 177, 185, 186, 187,191, 193, 194, 200, 201
Ordzenikidze 30
Oriental 235, 259, 277

Orion 240
Orion's Belt 18
Orkney 15
Our Future 3, 53

P

Paintings 8
Paper 229
Pavlopetri 59
Pentyrch 34, 78, 80
Periodic Table 226
Persians 213
Peru 40
Petroglyphs 97
Philanthropists 231
Phobos 31
Phoenix 10, 11, 26, 62
Pine Gap .. 46, 47, 66, 68, 69, 107, 114, 116, 240
Pinwheel 102
Pittsburg 216
Plasma 7, 9, 31
Plato 217
Pleiadan 235, 258, 259
Pleiades 273, 274
Pohnpei 42
Popocatepetl 37, 69
Portals 5, 37, 55, 56
President Jimmy Carter 238
President Donald Trump 234
President Dwight Eisenhower 21, 24, 26, 62, 66, 72, 207, 238
President George H W Bush 232
President Harry Truman 237
President Jack Kennedy 73, 74, 232
President Ronald Regan 232

Presidio 37, 237
Prince William Sound...................... 107
Printing 229
Protestant 222
Puerto Rico 37, 67, 68
Puma Puncu 18
Pumapunku 15
Pyramids 17, 18, 48, 214
Pyrrho 218
Pythagoras 217

Q

Quartz 15, 18
Quezalcoatl 16
Qui Shi Huang 16, 48, 219

R

RAF Bentwaters 31, 32
Railway Mania 223, 224
Railways 229
Rainhill Trials 223
Raleigh 222
Raphael 221
Renaissance 205, 221, 229
Rendlesham Forest 23, 31, 32, 33
Reptilians 128
Reptoids 26, 56
Reticulum 144, 145
Reverse Engineer.. 25, 52, 61, 74, 237, 238, 243
Richard Trevithick 223
Rio de Janeiro 67, 69
Robert Stephenson 223
Rocket 223
Rocket Fuel 244

Rocky Mountains 58
Romania 69
Roritzer... 222
Rose C 50
Rostov 236
Roswell................ 25, 61, 71, 236, 237

S

Sack of Rome 221
Salta 39
San Yungia 126
Saturn V 239
Schumann Cavity 241, 243
Schumann Resonance..................... 233
Sea Monster............................. 115, 170
Sea of Okhorsk 105, 157, 163
Sea-serpents 131, 142
Seismograph 229
Senganmori Mountain 42
Shag Harbour 28
Shakespeare 222
Shark 94
Siberia 17, 20, 79
Silk 229
Simonside Dwarfs 216
Skeptic 210, 211, 214, 218, 220
Skinwalker Ranch 5, 56, 65
Skulls 13
Sky-Train 129
Slave 218
Slave Race 263
Small Gray 21, 24, 26, 62, 66, 72, 107, 109, 112, 114, 123, 126, 128, 130, 135, 136, 139, 144, 146, 152, 154, 157, 159, 168, 187, 207, 216, 236, 245, 251, 253, 254, 257

Socrates 217
South America 3, 12, 13, 15, 38, 40
Sovereignty 72
Space Force 75, 76
Space Shuttle 76
Spacecraft 208, 234, 240, 252
SpaceX 240
Sphinx 48
Spy Planes 9
Squid 95, 115, 142
Stahl 225
Star Trek 214
Steam Train 223
Stephensville 11, 26
Stone Age 215, 218
Stonehenge 15, 213
Submarines 32, 42
Submersible vii
Sumarian 15
Sumeria 87, 119, 149, 209, 263
Syllabus 205, 235, 236, 245, 259

T

Taki Kyoto Maru 42
Tall Whites 236, 259
Tasmanian Tiger 114, 151
Tea 229
Telepathic 220, 235, 259, 273, 277, 281, 286
Telosians 92, 114, 140, 159
Tenerife 131, 196
Teotihuacan 18
Terengganu 118, 148, 177
Tesla 226, 227, 228, 232, 233, 234, 239, 241, 244
Tesla Coil 228, 232

Texas 11, 26, 38
The Nommo 90
Thylacine 114, 115
Tibet 49, 50, 64, 69
Time Machine 21
Tin. 215
Tinley Park 11, 26
Titans 16
Tokyo 27, 41
Toledo 216
TR3B 62, 75, 239
Trench 112, 113,126, 127, 146, 147, 157, 158, 187
Triangular 11, 12, 19, 26, 32, 33, 35, 36, 40, 48, 65
Trident 27
Tunguska 20, 29, 79, 241
Turkey 18, 59
Tutankhamun 15

U

Underground 85, 96, 97, 113, 120, 124, 129, 131, 139, 148, 150, 159, 207
Underwater .. 59, 66, 69, 85, 88, 91, 96, 97, 107, 110, 111, 113, 118, 122, 123, 127, 128, 130, 132, 139, 146, 153, 157, 159, 163, 178, 182, 199, 207
Urmahs (Felines) 115, 151
Usove 30
USS Eisenhower 21
USS Nimitz 23

V

V1 239

V2. 239
Valle of Lorestani 39
Valley of Death 17
Vandenberg 67
Varginha 40
Venezuela 39
Venus 9, 22
Vimanas 16, 64
Viracotcha 16
Volcano 37

W

Wales 10, 32, 34
Washington 22, 23, 63, 71
Westall High School 43
Westinghouse 227, 228, 231
Wheelbarrow 229
Wireless Power 236, 241
Witnesses. 8, 10, 12, 19, 27, 32, 35, 38, 44, 46, 47, 72
Wolf 222

Woodcut 8, 19
Wormhole 5, 57, 60

X

X-37B 76, 239
Xolotl (dog) 151

Y

Yeti 103, 113, 129, 130, 143, 162, 193, 251, 257
Yowie 114, 115, 117, 143, 174

Z

Zeno 218
Zeta Humans 144
Zeta Reticuli 285
Zhuangzi 219
Zimbabwe 10

REFERENCES

[1] The Alien Colonisation of Earth's Waterways P188 by Debbie Ziegelmeyer, Pub UnX Media USA 2021
[2] Celestial apparition over Nuremberg on April 14, 1561. Public Domain Image Archive.
[3] The Encyclopedia of Out of Place Artifacts p45 Martin Ettington 2022
[4] The Encyclopedia of Out of Place Artifacts p77 Martin Ettington 2022
[5] The Encyclopedia of Out of Place Artifacts p29 Martin Ettington 2022
[6] Lenticularis World Meteorological Association International Cloud Atlas
[7] Files released on 1974 'Welsh Roswell' https://www.bbc.co.uk/news/uk-wales-10863645
[8] UFOs: Few answers at rare US Congressional hearing https://www.bbc.co.uk/news/world-us-canada-61474201
[9] Broadhaven UFO 1977 by Justin Tulley 2024 Amazon
[10] UFOs Down Under Australasian Encounters by Barry Watts P154, pub Pegasus Education Group, Victoria, Australia 2017
[10] UFOs Down Under Australasian Encounters by Barry Watts P154, pub Pegasus Education Group, Victoria, Australia 2017
[11] UFO Insight - The Crestview Elementary School UFO Incident 1967
[12] IFL Science – The Aerial School Incident
[13] UFOs in US Airspace Hard Evidence P261 by John Scott Chace USA 2020
[14] New York Times https://www.nytimes.com/2020/04/28/us/pentagon-ufo-videos.html
[15] Blog of the American Heritage Centre – Discover History Flying Saucers. The papers of Jack Pickett 2022
[16] The Alien Colonisation of Earth's Waterways P36 by Debbie Ziegelmeyer, Pub UnX Media USA 2021
[17] Alien Contact: UFOs in European and Asian Air Space by John Scott Chace p112, Amazon 2020
[18] Something in the Sky – UFO Sightings in the UK p24 by Joshua Whittaker pub Austin Macaulay 2024
[19] Southern Illinois Tourism, Jackson County 2000
[20] Patch Tinley Park IL, UFO Sightings 2019
[21] Vice Newsletter by Nathaniel Janowitz 2023
[22] The Alien Colonisation of Earth's Waterways P184 by Debbie Ziegelmeyer, Pub UnX Media USA 2021

[23] Memory Cherish Palenque Astronaut
[24] SashaBlack Ancient Flier 2016
[25] Ancient Origins – Mysterious Phenomena 2013
[26] Ancient Origins – Mysterious Phenomena What became of the coneheads? By Karen Mutton 2019
[27] The Alien Agendas pp 11-21 by Richard Dolan. Pub Richard Dolan Press, New York 2020
[28] UNESCO World Heritage Convention: The List
[29] The Heritage Trust The Ishi-no-Hōden megalith theheritagetrust.wordpress.com/2014/10/27/
[30] UNESCO World Heritage Convention: The List
[31] Britannica
[32] Scientific & Esoteric Encyclopedia of UFOs, Aliens & Exterrestrial Gods V4 p156 by Maximillien de Lafayette pub New York 2014
[33] Sumerian Myths and Legends by Drayen Hallow Amazon 2024
[34] Egyptian Gods by Matt Clayton Pub Amazon 2020
[35] Inf news Did aliens help Qin Shihuang to rule the world?
[36] Ancient Origins the 'Bearded God' Named Quetzalcoatl - 0014066
[37] Ancient Code, Ancient Aliens, Viracotcha 2024
[38] Vimana: Flying Machines of the Ancients by David Childress Amazon 2013
[39] 101 Hindu Gods: Folklore and History with Pictures by I Chakraborty Amazon 2021
[40] Mahabharata – gathertales.com The Legend of the Battle of Kurushekta
[41] Rediscovering the lost city of Mohenjo Daro https://www.nationalgeographic.com/history/article/mohenjo-daro
[42] The Times of India Mar 22 2019
[43] Operation Disclosure Exopolitics: Ancient Alien Weapons
[44] The Giza Power Plant by Christopher Dunn. Amazon 1998
[45] The Giza Pyramids Alignment Guide by Czeszkiewcz Amazon 2022
[46] Britannica
[47] Britannica
[48] UNESCO World Heritage Convention: The List
[49] The culturetrip.com The Story Behind the Underground Cities of Turkey
[50] Nature NPJ Heritage Science 2018
[51] Milvian Bridge AD 312: Constantine's battle for Empire and Faith. Cowan & Brogain. Amazon 2016
[52] Something in the Sky. UFO Sightings across the UK p11 by Joshua Whittaker

Pub Austin Macaulay 2024
[53] The Soviet UFO Files P16 by Paul Stonehill Pub Bramley Books 1988
[54] Celestial apparition over Nuremberg on April 14, 1561
https://pdimagearchive.org/images/13e6b633-9551-48a1-8dc8-87f8139fdf97/
[55] State Museum of Berlin The air battle of Stralsund.
[56] Majic Eyes Only p48 by Ryan Wood pub Wood Enterprises USA 2024
[57] Majic Eyes Only p59 by Ryan Wood pub Wood Enterprises USA 2024
[58] Crystallinks – Project Haunebu
[59] Alien Contact: UFOs in European and Asian Air Space by John Scott Chace p262, Amazon 2020
[60] Alien Contact UFOs in European & Asian Air Space P22 Amazon John Scott Chace 2020
[61] UFOs in US Airspace Hard Evidence P377 by John Scott Chace USA 2020
[62] GIMBAL: Disclosure and the U.S.S. Nimitz y Topsy Kretz Amazon 2018
[63] Scientific & Esoteric Encyclopedia of UFOs, Aliens & Exterrestrial Gods V4 p56 by Maximillien de Lafayette pub New York 2014
[64] Encyclopedia of Alien Races p181 by Maesta Pastore Amazon
[65] UFOs in US Airspace Hard Evidence P377 by John Scott Chace USA 2020
[66] Encounter in Rendlesham Forest by Nick Pope, pub Thomas Dunne Books 2014
[67] GIMBAL: Disclosure and the U.S.S. Nimitz y Topsy Kretz Amazon 2018
[68] The Flatwoods UFO Monster The Skeptical Enquirer November 2000
[69] Shoot Them Down The Flying Saucer Air Wars of 1952 by Feschino 2000
[70] UFOs in US Airspace Hard Evidence P403 by John Scott Chace USA 2020
[71] UFOs in US Airspace Hard Evidence P403 by John Scott Chace USA 2020
[72] Daily Mail President Eisenhower had three secret meetings with aliens, former Pentagon consultant claims 16th Feb 2012
[73] The Extraterrestrial Species Almanac P25 Craig Compobasso, Red Wheel
[74] The Alien Colonisation of Earth's Waterways P184 by Debbie Ziegelmeyer, Pub UnX Media USA 2021
[75] The Alien Colonisation of Earth's Waterways P189 by Debbie Ziegelmeyer, Pub UnX Media USA 2021
[76] Trinity the Best Kept Secret by Vallee and Harris 2021
[77] UFOs in US Airspace Hard Evidence P195 by John Scott Chace USA 2020
[78] Majic Eyes Only p186 by Ryan Wood pub Wood Enterprises USA 2024
[79] Yesterday's America Area 51 History: Secrets Unveiled

[80] Alien Contact: UFOs in European and Asian Air Space by John Scott Chace p145, Amazon 2020
[81] Blog of the American Heritage Centre – Discover History Flying Saucers. The papers of Jack Pickett 2022
[82] Unsolved Mysteries Kecksberg UFO
[83] The Enduring Panic About Cow Mutilations. By Rachael Monroe. The New Yorker 2023
[84] The Alien Colonisation of Earth's Waterways P36 by Debbie Ziegelmeyer, Pub UnX Media USA 2021
[85] Alien Contact: UFOs in European and Asian Air Space by John Scott Chace p112, Amazon 2020
[86] UFOs in US Airspace Hard Evidence P261 by John Scott Chace USA 2020
[87] Vice Newsletter by Nathaniel Janowitz 2023
[88] Underground Alien Bases P15 by Jade Summers Self Published 2024
[89] UFOs in US Airspace Hard Evidence P283 by John Scott Chace USA 2020
[90] Malmstrom UFO Testimonials https://www.documentcloud.org/documents/9329-malmstrom-ufo-testimonials/
[91] Dark Files Bk1 P95 by Michael Schratt Amazon GB 2020
[92] The Alien Colonisation of Earth's Waterways P25 by Debbie Ziegelmeyer, Pub UnX Media USA 2021
[93] Trident missile test fails for second time in a row https://www.bbc.co.uk/news/uk-68355395
[94] Unidentified object' downed by U.S. fighter jets over Lake Huron PBS News Nation Feb 2023
[95] UFOs in US Airspace Hard Evidence P439 by John Scott Chace USA 2020
[96] UFOs in US Airspace Hard Evidence P439 by John Scott Chace USA 2020
[97] The Alien Colonisation of Earth's Waterways P28 by Debbie Ziegelmeyer, Pub UnX Media USA 2021
[98] The Soviet UFO Files P28 by Paul Stonehill Pub Bramley Books 1988
[99] The Soviet UFO Files P43 by Paul Stonehill Pub Bramley Books 1988
[100] The Soviet UFO Files P39 by Paul Stonehill Pub Bramley Books 1988
[101] The Soviet UFO Files P80 by Paul Stonehill Pub Bramley Books 1988
[102] The Soviet UFO Files P86 by Paul Stonehill Pub Bramley Books 1988
[103] The Soviet UFO Files P92 by Paul Stonehill Pub Bramley Books 1988
[104] The Soviet UFO Files P68 by Paul Stonehill Pub Bramley Books 1988
[105] The Soviet UFO Files P66 by Paul Stonehill Pub Bramley Books 1988
[106] UFO Insight 2017

[107] The Soviet UFO Files P70 by Paul Stonehill Pub Bramley Books 1988
[108] Something in the Sky. UFO Sightings across the UK p20 by Joshua Whittaker Pub Austin Macaulay 2024
[109] Broadhaven UFO 1977 by Justin Tulley 2024 Amazon
[110] The Dechmont Wood UFO Incident by Malcolm Robinson Amazon 2019
[111] The Bonnybridge (or Falkirk) Triangle ufos.ac.uk
[112] Encounter in Rendlesham Forest by Nick Pope, pub Thomas Dunne Books 2014
[113] Something in the Sky. UFO Sightings across the UK p24 by Joshua Whittaker Pub Austin Macaulay 2024
[114] BBC News Channel 20th June 2008
[115] https://herald.wales/national-news/pentyrch-the-greatest-ufo-cover-up-of-the-21st-century/
[116] Wales Online UK News March 2024
[117] Alien Contact: UFOs in European and Asian Air Space by John Scott Chace p59, Amazon 2020
[118] Alien Contact: UFOs in European and Asian Air Space by John Scott Chace p85, Amazon 2020
[119] Alien Contact: UFOs in European and Asian Air Space by John Scott Chace p85, Amazon 2020
[120] Alien Contact: UFOs in European and Asian Air Space by John Scott Chace p61, Amazon 2020
[121] Alien Contact: UFOs in European and Asian Air Space by John Scott Chace p112, Amazon 2020
[122] Nomanzone: UFO Sightings in Greece 2024
[123] Alien Contact: UFOs in European and Asian Air Space by John Scott Chace p80, Amazon 2020
[124] Sosacuelt.fi Mysteries in the sky
[125] The Alien Colonisation of Earth's Waterways P167 by Debbie Ziegelmeyer, Pub UnX Media USA 2021
[126] Stargates P98 by Betsey Lewis Pub Ingram Content group UK 2021
[127] The Alien Colonisation of Earth's Waterways P174 by Debbie Ziegelmeyer, Pub UnX Media USA 2021
[128] Alien Base page 313 by Timothy Good Pub Century London 1988
[129] Alien Base page 395 by Timothy Good Pub Century London 1988
[130] International Business Times 426515 2013
[131] Majic Eyes Only p278 by Ryan Wood pub Wood Enterprises USA 2024

[132] Unsolved Mysteries Mexico UFO
[133] Wired Mexican Airforce Film 2004-5
[134] UFOInsight UFOs Dr Enrico Bossa 2025 by Marcus Lowth
[135] Alien Base page 171 by Timothy Good Pub Century London 1988
[136] Alien Base page 171 by Timothy Good Pub Century London 1988
[137] Alien Base page 198 by Timothy Good Pub Century London 1988
[138] UFOs in Central And South America Airspace p169 by John Scott Chace 2020
[139] UFOs, Humanoids and Strange Phenomena of Argentina, Chile, Paraguay, Peru and Uruguay P48 by George Mitrovic
[140] UFO or pterodactyl over Argentinian lake? https://www.telegraph.co.uk/news/newstopics/howaboutthat/6166855/UFO-or-pterodactyl-over-Argentinian-lake.html
[141] New York Post Weird but True 2022 Aliens in Brazil
[142] Alien Base page 381 by Timothy Good Pub Century London 1988
[143] Media.com Christina Gomez Secret Argentinan UFO Cases 2024
[144] The Dragon's Triangle by Charles Berlitz pub Winwood Press New York 1989
[145] Alien Contact: UFOs in European and Asian Air Space by John Scott Chace p71, Amazon 2020
[146] Alien Contact: UFOs in European and Asian Air Space by John Scott Chace p217, Amazon 2020
[147] Alien Contact: UFOs in European and Asian Air Space by John Scott Chace p206, Amazon 2020
[148] Theufodatabase incidents The Kofu Incident
[149] The Dragon's Triangle by Charles Berlitz p32 pub Winwood Press New York 1989
[150] The Dragon's Triangle by Charles Berlitz p130 pub Winwood Press New York
[151] Britannica
[152] News on Japan article 145271
[153] UFOs Down Under Australasian Encounters by Barry Watts P81, pub Pegasus Education Group, Victoria, Australia 2017
[154] UFOs Down Under Australasian Encounters by Barry Watts P154, pub Pegasus Education Group, Victoria, Australia 2017
[155] UFOs Down Under Australasian Encounters by Barry Watts P57, pub Pegasus Education Group, Victoria, Australia 2017
[156] News.com.au April 27 2013
[157] Project Rainfall. The Secret History of Pine Gap by Tom Gilling pub Allen &

Unwin 2019
[158] UFO Insight Pine Gap
[159] UFO Insight Pine Gap
[160] UFO Insight Pine Gap
[161] World Press Tianjin Airport, 2024
[162] Aliens Amongst Us, The UFO Secrets of Africa by Adrienne Jaffery pub Ingram Content Group, UK 2024
[163] Aliens Amongst Us, The UFO Secrets of Africa by Adrienne Jaffery pub Ingram Content Group, UK 2024
[164] Alien Base page 30 by Timothy Good Pub Century London 1988
[165] Alien Base page 41 by Timothy Good Pub Century London 1988
[166] Alien Base page 57 by Timothy Good Pub Century London 1988
[167] Alien Base page 97 by Timothy Good Pub Century London 1988
[168] Alien Base page 100 by Timothy Good Pub Century London 1988
[169] Alien Base page 178 by Timothy Good Pub Century London 1988
[170] Mahabharata – gathertales.com The Legend of the Battle of Kurushekta
[171] Alien Encounter on the Moon: The Untold Story of Alien Contact During the Apollo Moon Landing by Robin Moore Amazon Kindle 2012
[172] Geographical Database of Bigfoot/Sasquash Sightings & Reports. https://www.bfro.net/gdb/
[173] Stargates P125 by Betsey Lewis pub Ingram Content group UK 2021
[174] BBC Travel 2022
[175] University of Nottingham: Pavlopetri 2013
[176] Harvard University Excavations at the Submerged Neolithic site of Atlit Yam, off the Carmel Coast of Israel
[177] Mahabalipuram Submerged Ruins 2022
[178] Göreme National Park and the Rock Sites of Cappadocia UNESCO World Heritage Center – The List
[179] The Alien Agendas pp 11-21 by Richard Dolan. Pub Richard Dolan Press, New York 2020
[180] USA Today 80470829007
[181] Crystallinks – Project Haunebu
[182] Trinity the Best Kept Secret by Vallee and Harris 2021
[183] UFOs in US Airspace Hard Evidence P195 by John Scott Chace USA 2020
[184] Yesterday's America Area 51 History: Secrets Unveiled
[185] The Day After Roswell by Corso & Birnes Pub Pocket Books 1997
[186] Blog of the American Heritage Centre – Discover History Flying Saucers

The papers of Jack Pickett 2022

[187] The Living Moon Particle Beam weapon 2004

[188] https://www.msn.com/en-au/news/other/top-secret-anti-gravity-spy-plane-tr3b-black-manta/vi-AA1w0LhM

[189] Encyclopaedia of Alien Races P181 Pub Amazon

[190] The Mount Musine UFO Encounters – What Attracts UFOs https://www.ufoinsight.com/ufos/close-encounters/mount-musine-ufo-encounters

[191] Stargates P27 by Betsey Lewis Pub Ingram Content group UK 2021

[192] Vocal.Media/History/Uncovering the Mystery of the Dark Pyramid

[193] Irish News UFO footage in Alaska 2024

[194] The Bonnybridge (or Falkirk) Triangle ufos.ac.uk

[195] Underground Alien Bases P15 by Jade Summers Self Published 2024

[196] UFO Insight Pine Gap

[197] Stargates P27 by Betsey Lewis 2021

[198] Alien Base page 224 by Timothy Good Pub Century London 1988

[199] Alien Base page 224 by Timothy Good Pub Century London 1988

[200] UFOs, Humanoids & Strange Phenomena of Central America, the Carribean and Mexico as well as the Atlantic Ocean P197 George Mitrovic

[201] The Alien Colonisation of Earth's Waterways P151 by Debbie Ziegelmeyer, Pub UnX Media USA 2021

[202] The Secret Alien Base of Bucegi Mountains by Aramescu Florin, UK

[203] The Book of Alien Races by Gil Carlson Blue Planet Press 2017

[204] The Alien Colonisation of Earth's Waterways P181 by Debbie Ziegelmeyer, Pub UnX Media USA 2021

[205] Forrestal Killed In 13-Story Leap www.nytimes.com/1949/05/23/archives

[206] Shoot Them Down The Flying Saucer Air Wars of 1952 by Feschino 2000

[207] The Plot to Kill President Kennedy in Chicago by Palamara Amazon 2024

[208] The Secret Space Program and Break-Away Civilisation by Richard Dolan P18

[209] The Secret Space Program and Break-Away Civilisation by Richard Dolan

[210] Cibernews:Tech 2023

[211] Blue Shuttle: The Manned Spaceflight Engineer Program spaceflighthistories.com

[212] National Defense Authorization Act For Fiscal Year 2020

[213] Boeing Defence Autonomous Systems

[214] The Soviet UFO Files P28 by Paul Stonehill Pub Bramley Books 1988

[215] The Alien Colonisation of Earth's Waterways P25 by Debbie Ziegelmeyer, Pub UnX Media USA 2021
[216] Celestial apparition over Nuremberg on April 14, 1561 https://pdimagearchive.org/images/13e6b633-9551-48a1-8dc8-87f8139fdf97/
[217] The Alien Colonisation of Earth's Waterways P25 by Debbie Ziegelmeyer, Pub UnX Media USA 2021
[218] https://herald.wales/national-news/pentyrch-the-greatest-ufo-cover-up-of-the-21st-century/
[219] UFO Friend or Foe? by Martin Thomas Pub 2025
[220] UNESCO World Heritage Convention. The List
[221] Scientific & Esoteric Encyclopaedia of UFOs, Aliens & Extraterrestrial Gods V4 p156 by Maximillien de Lafayette pub NY 2014
[222] The Alien Agendas pp 11-21 by Richard Dolan Pub Richard Dolan Press, New York 2020
[223] Sea Level in the Past 200,000 Years. University of New Orleans https://courses.ems.psu.edu/earth107/node/1496
[224] Alaska's Deadly Triangle p32 by Betsey Lewis 2023
[225] Stargates P27 by Betsey Lewis 2021
[226] UFO Sightings Hotspot 2024
[227] Irish News UFO footage of Alaska 2024
[228] International Business Times 426515 2013
[229] Oannes:The first extraterrestrial visitor in Antiquity by Carlos Allende https://saucerianbooks.blogspot.com/ 2022
[230] Nommo: The Celestial Beings Who Shaped Humanity by M M   Amazon Print Replica 2025
[231] Britannica
[232] Stargates P61 by Betsey Lewis 2021
[233] Encyclopaedia of Alien Races by Màesta Pastore
[234] The Naked Ape by Desmond Morris Pub Jonathan Cape London 1967
[235] Man and Dolphin: Adventures of a New Scientific Frontier Lilly, Dr John C. Lilly 1961 ·
[236] Nature Magazine 12 August 2015
[237] Consciousness of octopuses—on their own terms. Jenifer Mather *Animal Sentience 2025*
[238] Stargates P27 by Betsey Lewis 2021
[239] A History of USOs Vol1, Beginnings to 1969 by Richard Dolan Pub Richard

Dolan Press Rochester USA 2025

[240] The Secret Space Program and Breakaway Civilisation P48 by Richard Dolan Pub Richard Dolan Press Rochester USA 2025

[241] Ancient Aliens and the Exotic Gods by Bergen Amazon 2023

[242] Encyclopaedia of Alien Races by Màesta Pastore P151

[243] The Alien Agendas pp 14 by Richard Dolan Pub Richard Dolan Press, New York 2020

[244] UFOs, Humanoids and Strange Phenomena of Africa, Asia and the Middle East Geotge Mitrovic

[245] Russia's USO Secrets P111, Stonehill & Mantle Flying Disk Press 2020

[246] UFOs, Humanoids and Other Strange Phenomena of Russia. George Mitrovic

[247] The Alien Colonisation of Earth's Waters by Debbie Zieglmeyer Pub UnX Media Missouri 2021

[248] Russia's USO Secrets P111, Stonehill & Mantle Flying Disk Press 2020

[249] The Extraterrestrial Species Almanac p37 Craig Campobasso Red Wheel

[250] The Alien Colonisation of Earth's Waters by Debbie Zieglmeyer Pub UnX Media Missouri 2021

[251] Amazing Encounters with UFOs of West Coast North America, George Mitrovic

[252] The Alien Colonisation of Earth's Waters P163 by Debbie Zieglmeyer Pub UnX Media Missouri 2021

[253] Amazing Encounters with UFOs of West Coast North America, George Mitrovic

[254] The Alien Colonisation of Earth's Waters P175 by Debbie Zieglmeyer Pub UnX Media Missouri 2021

[255] A History of USOs Vol1, Beginnings to 1969 by Richard Dolan Pub Richard Dolan Press Rochester USA 2025

[256] UFOs, Humanoids and Strange Phenomena of Central America, the Caribbean and Mexico as well as the Atlantic Ocean, George Mitrovic.

[257] UFOs, Humanoids and Strange Phenomena of Argentina, Chile, Peru and Uruguay. George Mitrovic

[258] UFOs in Central and South American Air Spa. John Scott Chace 2020

[259] UFOs, Humanoids and Strange Phenomena of Argentina, Chile, Peru and Uruguay. George Mitrovic

[260] UFOs in Central and South American Air Space. John Scott Chace 2020

[261] Amazing Encounters with Monsters and Other Mysteries of Australia, New

[262] Amazing Encounters with Monsters and Other Mysteries of Australia, New Zealand, the Pacific and Antarctica by George Mitrovic
[263] UFOs Down Under by Barry Watts, Pegasus Education Group 2017
[264] A History of USOs Vol1, Beginnings to 1969 by Richard Dolan Pub Richard Dolan Press Rochester USA 2025
[265] Australia's Ancestral Cryptid: The Bunyip https://reactormag.com/australias-ancestral-cryptid-the-bunyip/
[266] Extinction of Thylacine https://www.nma.gov.au/defining-moments/resources/extinction-of-thylacine
[267] Blue Mountains Triangle. Rex & Heather Gilroy URU Publications 2006
[268] News.com.au April 27 2013
[269] Britannica
[270] News on Japan article 145271
[271] Russian USO Secrets by Stonehill & Mantle, Flying Disk Press 2020
[272] UFOs, Humanoids and Strange Phenomena in the Ukraine, Belarus, Moldavia and Scandanavia. George Mitrovic
[273] UFOs, Humanoids and Other Strange Phenomena of Russia. George Mitrovic
[274] Amazing Encounters with Monsters and Other Mysteries of Australia, New Zealand, the Pacific and Antarctica by George Mitrovic
[275] Amazing Encounters with UFOs of East Coast North America Vol 2, George Mitrovic
[276] Amazing Encounters with UFOs of Canada, George Mitrovic
[277] UFOs, Humanoids and Strange Phenomena of Central America, the Caribbean and Mexico as well as the Atlantic Ocean, George Mitrovic.
[278] Amazing Encounters with UFOs of East Coast North America Vol 1, George Mitrovic
[279] Amazing Encounters of the South. George Mitrovic
[280] UFOs, Humanoids and Strange Phenomena of Central America, the Caribbean and Mexico as well as the Atlantic Ocean, George Mitrovic.
[281] A History of USOs Vol1, Beginnings to 1969 by Richard Dolan Pub Richard Dolan Press Rochester USA 2025
[282] The Alien Colonisation of Earth's Waters by Debbie Zieglmeyer Pub UnX Media Missouri 2021
[283] The Alien Colonisation of Earth's Waters P151 by Debbie Zieglmeyer Pub UnX Media Missouri 2021

[284] UFO Contacts in Brazil, by Thiago Luiz Tictti. Flying Disk Press 2019
[285] Alien Base p 224 by Timothy Wood pub Century London 1988
[286] UFOs, Humanoids and Strange Phenomena of Argentina, Chile, Paraguay, Peru and Uruguay, George Mitrovic.
[287] Alien Base p 224 by Timothy Wood pub Century London 1988
[288] UFOs, Humanoids and Strange Phenomena of Africa, Asia and the Middle East Geotge Mitrovic
[289] UFOs, Humanoids and Strange Phenomena in the Ukraine, Belarus, Moldavia and Scandinavia. George Mitrovic
[290] UFOs, Humanoids and Strange Phenomena in the Ukraine, Belarus, Moldavia and Scandinavia. George Mitrovic
[291] UFOs, Humanoids and Other Strange Phenomena of Russia. George Mitrovic
[292] Russian USO Secrets by Stonehill & Mantle, Flying Disk Press 2020
[293] The Alien Colonisation of Earth's Waters P151 by Debbie Zieglmeyer Pub UnX Media Missouri 2021
[294] Stargates P27 by Betsey Lewis 2021
[295] The Mount Musine UFO Encounters – What Attracts UFOs To This Location? https://www.ufoinsight.com/ufos/close-encounters/mount-musine-ufo-encounters
[296] The Mount Musine UFO Encounters – What Attracts UFOs To This Location? https://www.ufoinsight.com/ufos/close-encounters/mount-musine-ufo-encounters
[297] UFOs, Humanoids and Strange Phenomena in Italy & Eastern Europe. George Mitrovic
[298] UFO Contacts In Italy Volume One
[299] UFO Investigations in Italy, Carlo Pirola Picadilly, NY City 2019
[300] The Extraterrestrial Species Almanac Craig Campobasso Red Wheel
[301] The Illustrated Guide to Reported Alien Species P25 David McDaniel 2017
[302] The Extraterrestrial Species Almanac p157 & 167 Craig Campobasso Red Wheel
[303] The Extraterrestrial Species Almanac p57 Craig Campobasso Red Wheel
[304] The Illustrated Guide to Reported Alien Species P51 David McDaniel 2017
[305] The Extraterrestrial Species Almanac p29 Craig Campobasso Red Wheel
[306] The Extraterrestrial Species Almanac p37 Craig Campobasso Red Wheel
[307] UFOs in US Airspace Hard Evidence P195 by John Scott Chace USA 2020

308 The Extraterrestrial Species Almanac p25 Craig Campobasso Red Wheel
309 The Illustrated Guide to Reported Alien Species P27 David McDaniel 2017
310 The Extraterrestrial Species Almanac P187 Craig Campobasso Red Wheel
311 UFO Friend or Foe? P38 by Martin Thomas Pub 2025
312 The Alien Agendas P175. By Richard Dolan. Richard Dolan Press 2020
313 UFOs, Humanoids and Strange Phenomena of Argentina, Chile, Paraguay, Peru and Uruguay, p10 George Mitrovic.
314 Encyclopedia of Alien Races P164 Maesta Pastore
315 UFOs and Aliens 2 p141 Carl Spartacus
316 UFO Friend or Foe? by Martin Thomas Pub 2025
317 UFOs and Aliens 2 p126 Carl Spartacus Dunstable
318 Alien Underground Bases Gil Carlson Wicked Wolf Press 2015
319 UFO Friend or Foe? by Martin Thomas Pub 2025
320 We Own 29% - ET Has the Rest by Martin Thomas. Pub 2025
321 We Own 29% - ET Has the Rest by Martin Thomas. Pub 2025
322 Letters From Mesopotamia A L Oppenheim Pub University of Chicago Press 1967
323 UFOs: Few answers at rare US Congressional hearing https://www.bbc.co.uk/news/world-us-canada-61474201
324 Files released on 1974 'Welsh Roswell' https://www.bbc.co.uk/news/uk-wales-10863645
325 The Order of the Dragon: The Battle Between "Other History and Accepted History by Colleen D Clements. Booksurge publishing 2006
326 The Order of the Dragon: The Battle Between "Other History and Accepted History by Colleen D Clements. Booksurge publishing 2006
327 A Humanized Version of Foxp2 Affects Cortico-Basal Ganglia Circuits in Mice by W Enard Cell Journal Volume 137, Issue 5p961-971May 29, 2009
328 Denisovan and Neanderthal archaic introgression differentially impacted the genetics of complex traits in modern populations. Dora Koller et al BMC Biology 2022.
329 Encyclopaedia of Alien Races by Màesta Pastore P151
330 Letters From Mesopotamia A L Oppenheim Pub University of Chicago Press 1967
331 Bird-Headed Deity, Denver Art Museum web site
332 Alien Contact: UFOs in European and Asian Air Space by John Scott Chace p71, Amazon 2020
333 Pyramids and Ziggurats – National Geographical Society Education

[334] The Chinese Pyramids and the Sun, AC Sparavigna · Torino Polytechnic 2012
[335] What's Inside the Pyramid at Chichén Itzá? Britannica, 2025
[336] Bronze Age Source of Tin Discovered The University of Chicago Chronicle 1994
[337] Bronze Age Source of Tin Discovered The University of Chicago Chronicle 1994
[338] Magic, Myth & Mystery of Dwarfs by Virginia Hagan Pub Cherry Lake USA 2019
[339] Iron Age, historical technological and cultural stage Brirannica 2025
[340] First evidence of crucible steel production in Medieval Anatolia Science Direct 2022
[341] Toledo, Spain, has been a sword-making hotbed for 2,500 years — now just 2 artisans are keeping the tradition alive. Business Insider 2021
[342] UFOs, Humanoids and Strange Phenomna of Andorra, Gibraltar, Spain & Portugal by George Mitrovic. Pub Dunstable, UK.
[343] Amazing Encounters with UFOs of East Coast America V2 by George Mitrovic Dunstable.
[344] The Duergar, The Dwarves of Simonside Hills. Spookyisles.com
[345] UFOs, Humanoids and Strange Phenomena of England by George Mitrovic.
[346] Bessemer Process Britannica 2024
[347] Hundred Schools Chinese History Britannica 2017
[348] Inf News/ history Did Aliens Help Qin Shi Huang to Rule the World?
[349] Giordano Bruno Italian philosopher Britannia 2025
[350] Richard Trevithick English Engineer Britannia 2024
[351] The Rainhill Trials by Anthony Dawson Pub 2018
[352] Deriving the Railway Mania. G Campbell Queens University Belfast Portal 2013
[353] George Hudson, the Railway King by Matthew Wells. Pub Pen& Sword 2024
[354] Social Support for Chemistry in Germany by Karl Hufbauer, University of California Press 1971
[355] Image & Reality by Alan Rocke Pub Amazon 2010
[356] Discovery of the Periodic Table by Matthew Lyons History Today 2021
[357] The Man Who Invented the Twentieth Century P26, by Robert Lomas. Headline Book Publishing 1999.
[358] Samuel F. B. Morse by Greg Timmons Biography>Famous Inventors>FamousPainters 2019

[359] Marconi's first radio broadcast made 125 years ago by Jonathan Holmes Pub BBC News/England/Local News/ Somerset 2022
[360] The Man Who Invented the Twentieth Century P261, by Robert Lomas. Headline Book Publishing 1999.
[361] Joe Kennedy: A Complex and Shocking Ambassador by Susan Ronald, Pub MacMillan 2021
[362] The First Hydro-Electric Power Plant in The World The Tesla society Website I
[363] The Man Who Invented the Twentieth Century P261, by Robert Lomas. Headline Book Publishing 1999.
[364] Nikola Tesla and the Tower That Became His 'Million Dollar Folly' by Gilbert King Smithsonian Magazine 2013
[365] UFOs, Humanoids and Strange Phenomena of Argentina, Chile, Paraguay and Uruguay by George Mitrovic. Pub Dunstable, UK
[366] UFOs, Humanoids and Strange Phenomena of Russia P34 by George Mitrovic Pub Dunstable UK
[367] Pentagon 'whistleblower' claims Vatican helped US retrieve UFO from Benito Mussolini New York Post June 13 2023
[368] The Day After Roswell by Corso & Birnes Amazon 2017
[369] Unidentified object' downed by U.S. fighter jets over Lake Huron PBS News Nation Feb 2023
[370] Magic Eyes Only p278 by Ryan Wood. Pub Wood Enterprises USA 2024
[371] Dreamland: An Autobiography by Bob Lazar Pub Amazon 2021
[372] The Alien Colonisation of Earth's Waterways P181 by Debbie Ziegelmeyer, Pub UnX Media USA 2021
[373] President Dwight D. Eisenhower's Farewell Address (1961) US National Archives
[374] The Secret Space Program and Break-Away Civilisation by Richard Dolan P18
[375] The Secret Space Program and Break-Away Civilisation by Richard Dolan
[376] Boeing Defense Autonomous Systems
[377] https://www.msn.com/en-au/news/other/top-secret-anti-gravity-spy-plane- tr3b-black-manta/vi-AA1w0LhM
[378] Cibernews: Tech 2023
[379] Spacecraft Type List Spacecraft.Fandom.com
[380] UFO Friend or Foe? P42 by Martin Thomas Pub 2025
[381] Blue MountainsTriangle by Rex & Heather Gilroy, URU Publications 2006

[382] The Dark Pyramid and Violent Nature www.imdb.com/title
[383] Pyramid in Alaska Can Power All of Canada? You-Tube: History 22 Aug 2023
[384] UFO Friend or Foe? P79 by Martin Thomas Pub 2025
[385] The Enigma of the Mars Pyramid by Boris Bigalke   Books on Demand 2024
[386] Pyramid Power by Toth & Nielson Destiny Books 1999.
[387] The Alien Colonisation of Earth's Waterways P184 by Debbie Ziegelmeyer, Pub UnX Media USA 2021
[388] UFOs, Humanoids and Strange Phenomena of Argentina, Chile, Paraguay and Uruguay by George Mitrovic. Pub Dunstable, UK
[389] We Own 29% - ET Has the Rest by Martin Thomas. Pub 2025
[390] We Own 29% - ET Has the Rest by Martin Thomas. Pub 2025
[391] Encyclopaedia of Alien Races by Maesta Pastore Pub Amazon
[392] UFOs & Aliens 2 by Carl Spartacus. Pub Dunstable UK
[393] Alien Races by Alan Fredrich 2021
[394] Alien Species Book by Neil Anami Pun Dunstable UK
[395] Illustrated Guide to Reported Alien Species by David McDaniel
[396] The Secret KGB Book of Alien Races Pub Dunstable UK 2025
[397] The Extraterrestrial Species Almanac by Craig Campobasso 2021
[398] The Alien Archive – The Ultimate Alien Database Jacob Sokol 2025
[399] UFOs, Humanoids and Strange Phenomena of Argentina, Chile, Paraguay and Uruguay by George Mitrovic. Pub Dunstable, UK
[400] The Extraterrestrial Species Almanac P47 by Craig Campobasso 2021
[401] UFOs, and Strange Phenomena of Austria, Belgium, Estonia, Germany, Holland, Kaliningrad, Latvia, Lithuania, Luxembourg, Poland and Switzerland p245. By George Mitrovich
[402] The Extraterrestrial Species Almanac P23 by Craig Campobasso 2021
[403] Encyclopaedia of Alien Races PP 146-163 by Maesta Pastore Pub Amazon
[404] The Natural History of the Dwarf by Richard Carrington Cosmo Books, Wem, England. Article from The Saturday Book, London 1958.
[405] The Natural History of the Dwarf by Richard Carrington Cosmo Books, Wem, England. Article from The Saturday Book, London 1958.
[406] The Hidden History of Elves & Dwarfs by Claude Lecouteux Pub Inner Traditions 2018
[407] The Origins of Wizards. Witches and Fairies. Simon Webb Pub Pen & Sword History 2022
[408] About Duende   duendedrama.org
[409] Magic, Myth & Mystery of Dwarfs by Virginia Hagan Pub Cherry Lake USA

2019
[410] We Own 29% - ET Has the Rest by Martin Thomas. Pub 2025
[411] UFOs, Humanoids & Strange Phenomena of Africa, Asia & the Middle East pp300-309 by George Mitrovic. Dunstable Uk
[412] The Alien Archive P123 – The Ultimate Alien Database Jacob Sokol 2025
[413] The Extraterrestrial Species Almanac pp101-109 by Craig Campobasso 2021
[414] The Extraterrestrial Species Almanac P103 by Craig Campobasso 2021
[415] Close Encounters of the Hawaii Kine. Hawaiian Airlines Issue 28-3
[416] Extraterrestrial Species Almanac P47 by Craig Campobasso 2021
[417] UFOs, Humanoids and Strange Phenomena of Austria, Belgium, Estonia, Germany, Holland, Kalingrad, Latvia, Lithuania, Luxembourg, Poland and Switzerland. P 245 (1988) By George Mitrovic Dunstable
[418] When Prophecy Fails by Festinger et. al. Pub Harper Torch 1956
[419] The Alien Archive P161 – The Ultimate Alien Database Jacob Sokol 2025
[420] Encyclopaedia of Alien Races PP 98 by Maesta Pastore Pub: Amazon
[421] Extraterrestrial Species Almanac P29 by Craig Campobasso 2021
[422] The Alien Archive P63 – The Ultimate Alien Database Jacob Sokol 2025
[423] UFOs in US Airspace Hard Evidence P403 by John Scott Chace USA 2020

www.ingramcontent.com/pod-product-compliance
Lightning Source LLC
Chambersburg PA
CBHW061214070526
44584CB00029B/3827